Aeronautical Decision-Making and Aviation Safety in the Alaskan Operational Setting

Aeronautical Decision-Making and Aviation Safety in the Alaskan Operational Setting introduces the reader to the real-life experiences of aviators who fly in remote settings such as Alaska in the United States. It covers the challenges related to limited aviation infrastructure and support that affect human factors like aeronautical decision-making and its impact on aviation safety.

Through a unique blend of meticulous case study analysis and semi-structured interviews with Alaskan pilots, this book offers a comprehensive understanding of the proverbial challenges of flying in Alaska. It uncovers the human factors elements specific to this environment, shedding light on the factors that influence a pilot's decision-making, which may contribute to the high rate of accidents in Alaska and other remote regions. The content is supported by historical and socioeconomic perspectives on remote-setting aviation operations. Global perspectives are discussed with narratives from one author's experiences flying to remote airstrips in Africa. The book concludes with practical recommendations to improve decision-making and aviation safety in these remote settings, making it a must-read for aviation professionals.

This insightful research is not just for academic consumption. It is a practical guide for aviation professionals, including pilots, dispatch teams, air traffic controllers, and aviation support personnel. It offers valuable insights into the human factors involved in flying in Alaska, which can be directly applied in other aviation resource-constrained geographical regions, making it an indispensable resource for those in the field.

Dana Atkins started her career flying medevac operations to remote locations across Alaska. She has flown for numerous operators in Alaska and previously flew the Boeing 767 domestically and abroad. She currently flies the Boeing 737 for a Legacy Airline in the United States. Dana holds a Master of Science

(MS) in Aviation from the University of North Dakota. Her main research interests are the human factors of bush and backcountry flying in remote settings.

Daniel Kwasi Adjekum is an assistant professor in the Department of Aviation, University of North Dakota, teaching courses in crew resource management, safety management systems, and human factors. He has over 24 years of experience as a former Ghana Air Force aircraft commander and aviation safety consultant. He has experience flying the BN-2T islanders and the Fokker F-27 into remote airstrips in Africa and previously flew the Boeing 737 aircraft for an international airline. Kwasi holds a Ph.D. in aerospace sciences. He is a globally recognized subject matter expert in aviation safety and a sought-after resource for global news media on aviation safety issues. He is also an International Air Transport Association (IATA) certified Safety Management Systems (SMS) implementation and control expert.

Aeronautical Decision-Making and Aviation Safety in the Alaskan Operational Setting

Dana Atkins and Daniel Kwasi Adjekum

CRC Press
Taylor & Francis Group
Boca Raton London New York

CRC Press is an imprint of the
Taylor & Francis Group, an **informa** business

First edition published 2025
by CRC Press
2385 NW Executive Center Drive, Suite 320, Boca Raton FL 33431

and by CRC Press
4 Park Square, Milton Park, Abingdon, Oxon, OX14 4RN

CRC Press is an imprint of Taylor & Francis Group, LLC

ISBN: 978-1-032-67474-2 (hbk)
ISBN: 978-1-032-67890-0 (pbk)
ISBN: 978-1-032-67886-3 (ebk)

DOI: 10.1201/9781032678863

Typeset in Times
by Apex CoVantage, LLC

Contents

1 Aviation Operations in Remote Settings

It is late morning on June 16, 2023, in a remote part of Alaska. Visibility is more than ten miles, and a pilot is loading his aircraft at a 750-foot airstrip situated atop a down-sloping, rock and grass-covered ridgeline near Shaktoolik, Alaska. He is flying a Title 14 Code of Federal Regulations (CFR) Part 135 on-demand charter flight supporting a remote bear hunting excursion. He loads a guide and some hunting gear into a Cessna 180H airplane and prepares for departure.

In the United States, Part 135 certificate holders can conduct a variety of operations: on-demand operations, which may include limited scheduled operations, or scheduled (commuter) operations, which allow unlimited scheduled operations as well as on-demand operations (FAA, 2023). On-demand operations can be conducted in airplanes with 30 or fewer passenger seats, a maximum payload capacity of 7,500 pounds, or in any rotorcraft. Commuter operations may be conducted in airplanes with a maximum passenger-seating configuration of nine seats and a maximum payload capacity of 7,500 pounds (FAA, 2023).

Therefore, having this operator use the Cessna 180*H* for his operations was unsurprising. The Cessna 180 has been a popular aircraft type in Alaska for decades. It is a robust aircraft, capable of landing on wheels, floats, or skis, and its performance capabilities allow it to maneuver into a wide range of landing zones. Modifications to the airframe and engine are standard and help enhance performance; the Cessna 180 was originally certified with the Continental $O-470$ engine. This is a six-cylinder, horizontally opposed, carbureted engine rated 230 horsepower. It is an all-metal aircraft capable of carrying four to six passengers with a useful load of approximately 1,200,

DOI: 10.1201/9781032678863-1

depending on the model and modifications (AOPA, 2024). For the day's operation, the Cessna 180 was a solid choice of aircraft.

About an hour earlier, the Part 135 pilot had uneventfully departed the remote off-airport airstrip with two hunters/clients. Their hunting expedition had ended, and they were being chartered from the mountain ridgeline airstrip back to Unalakleet Airport (PAUN) with their gear. The pilot returned to the airstrip to pick up the two remaining hunting guides. They waited near their camp, not far from the landing zone. During that hour, the two guides waited for the pilot to return, and the winds seemed to increase in intensity—it was becoming gusty and changing direction. At this remote airstrip, pilots normally landed uphill on about a 060° heading, then departed in the opposite direction, downhill, on about a 240° heading.

The pilot had flown in and out of this airstrip numerous times. The guides observed that after takeoff, the airplane would dip out of sight below terrain off the departure end, climb back into view, and climb out of the valley. The pilot finishes loading more gear and boards one of the guides. He then starts the engine and taxis to the end of the strip for takeoff.

The pilot initiates the downhill takeoff and lifts off around 11:35 Alaska time. The aircraft struggles to gain altitude, resulting in a collision between part of the aircraft's wings and a small stand of trees about two-thirds down the down-sloping runway. The impact causes a loss of control, and the airplane impacts the tundra in a steep nose-down, wings-level attitude about 1,200 ft. from the damaged trees.

The second guide, who was to be picked up on the last flight of the day, watched the initial portion of the takeoff roll. Nothing seems abnormal, so he turns away and does not watch the rest of the takeoff. When he does not hear the familiar sound of the aircraft climbing out above terrain, he looks. He cannot see the aircraft climb out, so he runs to the edge of the ridgeline. The aircraft had crashed about 300 feet below the strip. This guide sends an SOS message from his GPS tracker and hikes to the wreckage to check for survivors.

The damage to the airplane is substantial. The right wing is crushed along its span, and both wings have been displaced and crushed by impact with the terrain. The engine is separated from the airframe and rests inverted behind the left wing. The propeller hub is fractured, and the blades are separated. The pilot and passenger are fatally injured. Figure 1.1 shows a photograph of the wreckage.

Unfortunately, this narrative is based on a factual accident in 2023 (FAA, 2023) and highlights the operational and safety risks for flight operations in remote settings such as Alaska. Remote settings present a multitude of challenges for aviation service providers and operators. These remote operational environments often lack vital logistical and operational infrastructure, such as airports with structured runways and handling facilities, advanced surveillance

FIGURE 1.1 Wreckage of the mishap aircraft. A red C180 sits partially inverted in the grass. The major components, like the wings, tail, and landing gear, are mostly crushed

Source: NTSB, 2023.

and navigation aids, real-time weather service provisions, and maintenance facilities. The airspace also lacks vital air traffic management services such as flight following and radar monitoring. Operations that intentionally take place off-airport are even more susceptible to the hazards of remote environments.

These remote settings are also rife with adverse weather conditions like extreme cold or heat, high winds, low visibility, and other obscurations that further complicate operations. The topography can be daunting—extreme highlands, tropical forests, tundra, and desert-like conditions are standard features of remote locations. The big question *is: Why conduct flight operations in these places if the safety risk is obvious?* The reality is that these aviation operations support economic, scientific, tourism, and humanitarian end-states and need to be sustained through effective safety risk management.

So why use Alaska as a basis for studying remote settings? Alaska has varied terrain features and population centers that represent other locations globally. More importantly, data is available to study, and resources are accessible for further research. This book explores the nuanced flight operations in such remote settings and identifies some challenges and threats to safe flight operations in Alaska.

We provide the reader with a firsthand look at various operations in Alaska through a review of extant literature and a synthesis of interviews conducted with aviation operators (namely pilots) to understand their lived experiences flying in Alaska (phenomenological approach). We also explore the perspectives of these operators on various human factors constructs such as pilot decision-making and biases, communication, team dynamics, social norms, and pressure and safety management that impact safe flight operations. We then provide the reader with practical recommendations for continuous monitoring and improvements in operational safety based on the feedback from these critical stakeholders.

WHAT ARE REMOTE SETTINGS?

To put it simply, remote settings are locations where accessibility is difficult. You may think of remote settings and rural areas near where you live. You may think back to movies or media you have seen that show extreme environments and difficult living conditions. To understand remote settings, it helps to first think of their characteristics:

- Lengthy travel time or travel via different transport models
- Small population sizes
- Lack of access to essential services such as fire and rescue, police, hospitals, or timely mail services
- Geographic location or terrain features that impede access
- Extremes in weather conditions such as temperature, storms, or niche weather phenomena

In many cases, remote settings are removed from larger population centers and urban areas due to geographic features that physically separate a remote place from somewhere more populated. The difficulty of a remote setting is the combination of *many* challenging features and how they interact to restrict accessibility. These features could be a variety of the following:

- Bodies of water
- Mountainous terrain, including entire mountain ranges
- Inhospitable swathes of land, such as tundra, desert, wetland, and so forth
- Landlocked locations, or conversely, islands not accessible via road
- Protected lands (Federal or State) that limit infrastructure development

Colloquially, remote settings are often called the "bush" or the "back-country." It is reasonable to suggest that there is a wide range of these remote

FIGURE 1.2 A United Nations Humanitarian Assistance Services Cessna Caravan at Tshikapa Airstrip in the Democratic Republic of Congo

Source: WFP/Elizabeth Byrant.

settings across the globe. Tropical forest areas of Africa, tundra in the Canadian north, desert in the Middle East, and the Outback of Australia are remote locations with vastly different terrain, weather, and population features, which is a small sampling (Michalski & Bearman, 2014). Figure 1.2 shows an example of an aviation operation in a tropical climate. Seasonal changes to the topography may also hinder access—for example, travel over an ice road in winter months may replace boat or water access in summertime. Air travel is a consistently reliable means of transportation throughout the year, yet it comes with higher costs to consumers and different logistical challenges. While remote locations may have vastly different geographic features and cultural customs, accessing communities in rural settings poses similar challenges globally.

TRANSPORTATION IN REMOTE SETTINGS

Public transportation is essential for daily life in urban communities. It provides flexibility and access to resources and social activities without significant personal economic impact. Many different transportation modalities create access within these urban communities; however, road and rail systems (including buses and metro/subway systems) are typically the most accessible for the general population.

In remote settings, however, infrastructure development for public transportation may be limited partly due to physical constraints such as geography and topography. These geographical elements impose higher construction and maintenance costs that are only justified by higher economic and social opportunities the infrastructure may provide. Areas with low physical barriers tend to have transportation networks with heightened network density and connectivity, which underlines the geographical influence on mobility and productivity. Climate constraints and weather disruptions also lead to high costs for transportation infrastructure development (Kuklina et al., 2022).

Some researchers suggest that complex socioeconomic factors limit infrastructure development in remote settings. Kuklina et al. (2022) suggest a misunderstanding of the value of access to modern infrastructure (especially transportation) and the associated costs. Interestingly, the authors infer that a developed transport infrastructure in remote regions can contribute to its economic potential.

Generally, transportation infrastructures tend to be capital-intensive, and limited fund availability can hinder their development or maintenance. When resources are scarce, it is essential to carefully consider the most appropriate allocation to reap the best economic and social benefits (Rodrigue, 2024). Infrastructure development could be impaired if these benefits are uncertain, particularly in remote settings. The reality is that some of these remote settings do not have vibrant economic activities to offset the significant investments required to improve transportation infrastructure.

In large-scale remote locations, like the tundra and Arctic regions of Alaska and Canada, concerns have been raised about the lack of resource allocation for transportation infrastructural improvements. This invariably affects social and economic mobility (Grabowski et al., 2016; Lee, 2023).

Without cost-effective and reliable transportation networks, education, health services, and employment opportunities may suffer (Horn et al., 2021). Disaster response, emergency management, and search and rescue operations are also limited. For example, the State of Alaska State Emergency Operations Center (SEOC) cites high financial and technological resource costs in its 2024 Emergency Operations Plan. They anticipate that relief efforts in remote parts of the state will be challenging and outcomes uncertain (State of Alaska DHS, 2024). Figure 1.3 shows humanitarian aviation missions in a global setting.

Because residents of remote settings may not have the benefit of mass public transportation, they need to rely on other means to move about their communities. Inhabitants of remote communities must reconcile the difficulties of living in such austere locations with the personal benefits they get. As such, the population in these places tends to be small. Residents may have heritage in the area or belong to the ancestral communities that initially inhabited the regions.

FIGURE 1.3 Mission Aviation Fellowship Kodiak (left) and Cessna 208 Caravan (Right) undertaking mission and humanitarian flights to the remote village of Soba, Indonesia

Source: Used with permission from Mission Aviation Fellowship; photo by Mark and Kelly Hewes.

In recent times, there has been a paradigm shift in rural communities. Most rely on the same goods and services that urban dwellers do. Residents travel to larger population centers to access medical care and participate in social functions. Remote communities may have subsistence lifestyles, while others rely heavily on imported goods. When land-based transportation models are either underdeveloped or non-existent, aircraft usage becomes essential for these communities (Foster, 1990; Anderson, 2000).

REMOTE AVIATION OPERATIONS (BUSH FLYING)

So now that we have defined remote settings, what does operating an aircraft in the bush mean? The term "bush" has been used in Canada and Alaska for many years to describe locations inaccessible by the road or ferry network. Bush flying was born between WWI and WWII as commercial aviation became popular throughout the United States and Canada. Pilots returning from WWI began to use aircraft to begin exploration for mineral development and assist in transportation for early mountaineers and alpinists (Ringsmuth, 2015). Since then, bush flying has expanded to include medical evacuation flights, mail contract fulfillment, air tours, and transporting people and goods. Because the airplane is the most reliable means of transportation to and from remote locations, it has become an essential part of life.

Geography of the Bush

Geographical features could be the single most impactful factor in bush flying. Globally, there are varied features that make each remote location unique. Geographical features can impact weather patterns, which impact flight operations and make aeronautical decision-making more nuanced. "Extreme" is a word commonly used to describe the backcountry. Extreme terrain, weather, and airports are part of the draw for some pilots to operate here. Navigating mountainous terrain requires training and precision (Pocock, 2017). In other areas, barren land provides no relief from sun and wind elements. Any features like these, combined with the sparseness of ground-based navigation aids, require further training and experience to operate safely.

Aircraft

Many aircraft used in the bush (over 85%) are single-engine and range from small two-place Piper Super Cubs to larger Cessna Caravans (Conway et al., 2004; FAA, 2021a). Other aircraft types frequently used in bush operations are no longer in production (like the DeHavilland Beaver and the Cessna 180). This contributes to the phenomenon of a declining fleet—when aircraft become permanently out of service (due to lack of parts or substantial damage), companies must decide how to proceed when acquiring a replacement aircraft, which could be cost-prohibitive or impossible.

Multi-engine aircraft are also used in bush or backcountry environments. These aircraft types have the added benefit of increased range, cargo capacity, and reliability that comes with two engines. However, operating costs are increased due to more sophisticated maintenance, fuel consumption, and the associated financial ramifications of hiring, retaining, and training pilots certified to fly them.

Aircraft used for bush operations are frequently modified to withstand harsher operating environments. For example, during summertime, replacing wheels with floats allows pilots to operate in remote parts of the backcountry.

FIGURE 1.4 A Piper Super-Cub ski plane rests on a frozen, snow-covered lake

Source: Dana Atkins, 2020.

It facilitates landings on bodies of water such as rivers, lakes, or oceans. In the winter, retrofitting aircraft with skis (such as the example in Figure 1.4) can open terrain inaccessible in the summertime, as ski strips can be put anywhere the terrain is relatively flat. These operators can access remote places where land-based airports may not be feasible.

Depending on where an operation is located geographically, it may only be possible to operate in these configurations seasonally, which creates additional challenges for pilots to maintain proficiency throughout a calendar year. The skill set required to operate on either floats or skis is also heightened—float flying requires an additional rating for pilots and the added element of being able to read and interpret water and understand the effect of wind on the landing surface. While no additional rating is required to operate on skis, novel challenges are associated with these operations. Exposure to the elements, the risk of getting stuck in deep or unstable snow, and challenges associated with freeze/thaw cycles make ski flying difficult to execute safely. Ski flying is predominant primarily in extreme northern and southern latitudes like Alaska, Canada, and the Antarctic (FAA, 2004).

From an operational perspective, floatplanes frequently have less crosswind component capability than their land plane counterparts. Directional control can be more demanding on water because the surface yields more and has less surface friction than on land, and seaplanes lack brakes. These factors increase the floatplane's tendency to weathervane into the wind (FAA, 2004). There may be periods when a pilot cannot land directly into the wind due to water traffic in the area, obstructions on or under the water, or a confined landing area, such as a river or canal.

Water landings with crosswind components require the same techniques (crabbing) as landing a wheel plane to minimize sideways drift during touchdown and maintain directional control afterward. If a floatplane touches down while drifting sideways, the sudden resistance as the float contacts the water can create a skidding force. A combination of the skidding force, wind, and "weathervaning" as the floatplane slows can lead to a loss of directional control, with the potential for the wingtip to contact water and the aircraft to flip over (FAA, 2004).

Title 14 of the Code of Federal Regulations (14 CFR) Part 61 does not require specific pilot training and authorization to operate ski planes; however, the FAA strongly recommends that pilots train with a qualified ski plane flight instructor (FAA, 2004). Skiplane also requires dexterity to operate in glaciers, tundra, or frozen lakes. The FAA recommends evaluating each new landing site thoroughly, obtaining advice from well-qualified pilots already familiar with the area, and staying within the limitations of personal skill and experience. These can contribute to safety and reduce risks.

For off-site airport operations, when pilots arrive at a prospective landing site, a pass should be made over the landing area to determine the landing direction and quality of the surface and if a safe approach and landing can be completed. Ski planes can be clumsy and less maneuverable while on the ground. They tend to weathervane with the wind and need considerable space to maneuver. Maneuvering on the ground and parking requires special techniques acquired only through practice.

Pilot Group/Experience

In bush flying, experience comes with time and exposure. Some young pilots are often eager to hone their aeronautical skills and decision-making capabilities through backcountry flying. Many fledgling pilots (see a picture of the author in figure 1.5) fly in remote locations to "cut their teeth" and build flight time for their careers in other parts of the industry (often legacy or major airlines). This type of experience can be precious to a pilot—they undoubtedly learn great aircraft control (i.e., stick and rudder skills), solve problems in challenging scenarios, and develop a broad range of experiences flying into and out of various airports with limited navigational facilities. These pilots benefit from mentoring, shared experience, and opportunities to fly with pilots experienced in these areas. This type of relationship is necessary to create a solid safety culture.

The aviation industry is cyclical, and pilot hiring is the same. The airlines are the most prominent subset of the industry, and as a result, their hiring cycles are vital drivers of pilot supply and demand. When there is a demand for pilots throughout the industry, it is common to see many pilots leave positions in the bush to favor positions with the airlines. With a nominally high turnover of this pilot group, experienced pilots who opt for a lifestyle flying in remote settings bring invaluable organizational skills, especially in retaining institutional knowledge.

In North America, the community of bush pilots has historically been predominately male. However, there have been some female pilots within this community, and one is worth mentioning. Ellen Evak Paneok was the first Native woman to become a licensed pilot in Alaska. She started as a career bush pilot flying cargo and mail for Barrow Air, Inc. She used her Piper Cherokee Six to travel to the remote communities near Barrow, Alaska. She was a skilled pilot, respected for her ability to fly in the challenging environment of Western Alaska (Ruffing & Cochrane, 2023).

Another remarkable bush pilot was Beryl Markham, who trained to become Africa's first female professional pilot. In the 1930s, Beryl flew to remote mining towns in Eastern Africa, delivering mail and medicine

FIGURE 1.5 Photo of the author (Dana) standing in front of a B200 King Air in 8 inches of snow after a Part 135 medevac flight operation

Source: Dana Atkins, 2018.

to settlers. She did her aircraft maintenance (a valuable trait for early bush pioneers) and frequently navigated with only the guidance of a compass and the stars. At the time, Africa was primarily unmapped, and there were no navigational aids to direct her. She was also the first pilot to fly across the Atlantic from East to West, a remarkable feat for 1936 (Ross, 2015).

When you think of the type of person who lives or works in remote settings, it is natural to assume that they are a rugged character. Bush pilots have long been characterized by their willingness to take risks and work hard. They portray what has become known as a "cowboy" attitude toward flying. Cowboy refers to their approach to flying, attitudes, and disposition toward their job. Is this a fair assessment? We will investigate how attitude and personality play into pilot traits and hear from bush pilots for their take on the cowboy persona.

Airports/Aerodromes/Airfield Infrastructure

In remote areas, airports exhibit considerable variability in their location and condition. Often, runways are unimproved and may lack regular maintenance or upkeep. Runway surfaces range from grass or gravel to dirt or pavement. Even in the case of paved runways, modern amenities like surface markings or lighting may be absent. Daily maintenance such as snow plowing, sweeping of foreign object debris (FOD), and runway condition reports are frequently unavailable due to limited infrastructure and personnel.

Bush flying also involves operations in off-airport locations where pilots must assess the terrain—land, water, or snow—to determine if it is suitable for landing. When used frequently, these areas are referred to as "bush strips," though they may be popular landing zones among pilots, they lack official airport status. For example, a typical off-airport location operation can involve landing on gravel strips, dirt/mud strips, glaciers, frozen lakes/rivers, and tundra, requiring careful planning and risk assessment. Landing in glacier areas with very steep slope angles requires evaluating the possibility of avalanches. Pilots are advised not to land near the bottom of a valley because ice falls may exist and provide rough and unusable terrain. Figure 1.6 shows a Piper Super Cub (PA18–160) parked on the ice at a Ruth Glacier, Central Alaskan Range site by Tundra Pilot photographer Wayne Mushrush in 2022. The strip type is a glacier with rough textured ice and 800 feet long at an elevation of 2,634 feet.

Clear ice can be incredibly slick under certain conditions for aircraft maneuvering in frozen lakes and rivers. Once the aerodynamic controls become ineffective due to the loss of airflow, directional control may not be possible. Tundra typically comprises small clumps of grass that can support snow and make ridge lines invisible. They also hide obstacles and obscure

FIGURE 1.6 Piper Cub (PA18-160) parked at a location in Ruth Glacier in the Central Alaska Range, elevation: 2,634 feet, coordinates 62.821, −150.619

Source: Photo used with permission. Photographer: Wayne Mushrush; URL: https://tundrapilot.com.

holes that may be too weak to support an airplane. Other hazards may be surface inundations, wildlife, the length of the landing strip, and features that could create confinement after landing.

The FAA Seaplane/Floatplane/Skiplane Operations Handbook 8083-H-23 (2004) provides some guidance for landing at such off-site locations. Details such as the ability to conduct a safe approach/departure, aborted landing plan, wind conditions, assessment of surface conditions, identification of a touchdown spot, and measurement of the usable length of the strip or landing area are critical. The Handbook recommends a low pass over such off-site strips to closely examine the surface conditions before committing to land.

Pilots should plan and configure for a soft-field landing with a stable approach for the low pass. Then, execute a gentle soft-field touchdown, controlled with power, while remaining near takeoff speed for approximately 600 *to* 800 feet, and initiate a go-around.

The final landing pattern should factor in the minimum obstacle clearances and prominent vertical obstacles. Landing approaches may require non-standard flight circuit patterns around the terrain and a relatively slower approach speed when operating in confined areas, such as between mountains, to land on a plateau strip. Pilots should identify their touchdown spots and visual descent points for any aborted landing while on the approach and have a go-around/escape plan in case of changes in wind conditions or wildlife hazards crossing the strip during landing.

Outside of choosing a safe landing zone, flight operations require pilots to make informed and safe decisions about whether to fly (go or no-go decisions) and in-flight navigation. Understanding the weather is a significant component of this. The geography of bush locations can contribute to extreme weather and climate patterns that are difficult to predict or forecast.

Moreover, not all destinations have reliable weather reporting facilities or forecast products. For pilots who rely on onboard and ground-based navigational equipment to fly safely in instrument meteorological conditions (IMC), the lack of regular and updated weather reports such as the Meteorological Aerodrome Report (METAR) can limit safety.

METAR provides current weather observations at airports, such as wind speed and direction, visibility, cloud cover and height, temperature, dew point, precipitation, and barometric pressure (ICAO, 2011). Another essential weather product not readily available at remote locations is a Terminal Aerodrome Forecast (TAF), which acts like a meteorological crystal ball. While METAR provides current conditions, TAF provides pilots with a weather outlook for the next 24–30 hours and highlights crucial details like expected cloud layers, visibility, wind shifts, precipitation, and thunderstorm probabilities (FAA, 2023). These products aid pilots in determining their suitability for flight.

Pilots flying into these airports must consider the prevailing and anticipated weather conditions in their decision-making process. They must also consider regulatory requirements and legal restrictions dictated by weather conditions when operating at that airport. Weather conditions influence crucial factors such as the feasibility of initiating an instrument approach to specific runways or selecting alternate airports. For instance, in Alaska, some airports have instrument approaches but lack the necessary weather-reporting capabilities to conduct an approach legally. For example, Central Alaska has two RNAV instrument approaches to its runways but relies on weather from Fort Yukon, an airport over 60 nm away (AASP, 2017).

Pilots rely on information from various sources—the internet, other pilots, people on the ground—and the information getting to each pilot must be reliable, factual, and timely. Where available, accessing current weather reports may be difficult, as many locations in remote areas do not have internet or cellular network signals. Training weather observers at outstations, enabling devices with better cellular network reception, and increasing the accessibility of text-based weather products all help improve these resources for pilots.

As mentioned earlier, one of the challenges faced by operators flying under instrument flight rules is the occasional technical obsolescence of ground-based navigational and surveillance aids. The terrain and weather conditions make routine maintenance and systems upgrades laborious, time-consuming, and expensive. The cost-benefit of consistent technological improvements to facilities at some of these remote settings diminishes, and there are scenarios where older technologies are still maintained and used until attrition.

For example, in a 2021 report sponsored by the FAA on Alaska aviation safety, it was noted that there are communication and surveillance gaps associated with Very High Frequency (VHF) Omnidirectional Radio Range (VOR), Distance Measuring Equipment (DME), and Non-directional Beacon (NDB) airways (FAA, 2021b). These navigation systems rely on ground-based facilities (FAA, 2021b). With the increasing cost of upkeep and maintenance, some of these aids fall into disrepair. This presents twofold problems: pilot training and facility reliability:

1. Training: Some pilots fly aircraft that still have onboard technologies such as VOR, DME, and NDB receivers. These systems depend on ground-based navigational aids. In the mainstream aviation operation system, GPS has replaced many ground-based systems. If these operators are not proficiently trained to understand the limitations of ground-based systems, operational errors that adversely affect flight safety can be made.

   In the United States, pilots earn their initial instrument ratings using the instrumentation in the aircraft where they take check rides. Pilots do not necessarily need to demonstrate proficiency in its use if the aircraft does not contain particular instrumentation. However, during the practical test, the applicant must demonstrate appropriate knowledge, risk management, and skills related to onboard instruments.

   Let us look at this quote from the FAA Airman Certification Standards (ACS) for instrument rating practical tests to give us some context:

   To assist in aircraft management during the practical test, the applicant must demonstrate automation management skills by utilizing installed equipment such as autopilot, avionics and systems displays, and a flight management system (FMS). The evaluator is expected to test the applicant's knowledge of the installed and operative systems during the practical test's oral and flight portions (FAA, 2018).

Training standards vary company by company depending on the ruleset each operator follows. Regardless, pilots' skills get rusty when they are not utilized.

2. Reliability: As technology ages, it is replaced with more advanced alternatives. Regrettably, this means that the older systems (which now may be considered obsolete) fall into disrepair or are decommissioned altogether. Older systems tend to have more frequent maintenance problems, leading to more downtime while maintenance is performed. This reduced reliability results in less usability for pilots and operators in the bush.

   There is continuing advancement in the dependability and accuracy of space satellite-based navigational systems. The global position systems (GPS), with features such as Wide Area Augmentation Systems (WAAS) and Automatic Dependent Surveillance-Broadcast (ADS-B) systems, have become critical components of essential communication, navigation, and surveillance (CNS) in the National Air Space (NAS) and flight safety/control systems. Additionally, some satellite communications avionics use GPS signals for operations in remote airspaces. It is ADS-B's sole aircraft position-reporting source (FAA, 2024).

GPS is also a necessary component of the Aircraft Terrain Awareness and Warning System (TAWS)—an aircraft warning system that provides cautionary alerts for impending terrain. These new technologies have gradually diminished the dependence on terrestrial-based navigation systems such as VOR, DME, and ADF in bush flying operations. To safely use GPS-based precision and non-precision approaches to their full capabilities, reliable weather data, pilot training, and equipment onboard aircraft capabilities are still needed.

Interestingly, many aviation accidents and incidents in remote locations stem from inadvertent flight from visual flight rules (VFR) into instrument meteorological conditions (IMC). This transition can lead to pilot spatial disorientation and loss of control inflight (LOCI). Frequently, controlled flight into terrain (CFIT) is the final cause of this flight path. A U.S. National Transportation Safety Board (NTSB) Report suggests that CFIT, LOCI, midair collision, and unintended encounters with IMC comprised about 80% of fatal Part 135 accidents in Alaska from 2008 to August 2019 (NTSB, 2020). Increasing the availability and reliability of ground-based weather resources and forecasting tools could help mitigate the number of CFIT accidents in these complex aviation operational settings.

Regulatory Challenges

Due to the remote nature of bush flying operations and how distal they may be from established regulatory structures, effective safety regulatory oversight can be a challenge. For example, as part of the FAA safety initiative for Alaska, stakeholders expressed concerns about the high turnover among principal operations inspectors (POIs). This turnover leads to a lack of continuity, limited communications, and potential gaps in regulatory oversight between operators and Flight Standards (FAA, 2021b). Stakeholders strongly recommended analyzing turnover rates, including staff responsible for interacting with the public, such as flight standards representatives.

In the United States, Federal Aviation Regulations (FAR) Part 135 safety has been on the NTSB's Most Wanted List for improving aviation safety for multiple years. The NTSB has recommended significant safety improvements through CFIT-avoidance training, adopting safety management systems (SMS), and flight data monitoring programs for Part 135 operators (NTSB, 2019). Unfortunately, the NTSB (which investigates aviation accidents and incidents) can only issue recommendations to the FAA (which regulates the industry).

This chasm of responsibility means that recommendations are frequently left unaddressed. Another critical challenge with operational safety oversight in the Alaska region has been the carriage of passengers by operators serving seasonal lodges, which do not fall under Part 135 regulations. There have been suggestions that such transport operations should be transferred to Part 135 operations since they fly passengers (FAA, 2021b).

Fly Alaska *Remote Setting Operational Factors and Safety Implications* 2

The state of Alaska epitomizes the "bush" in the United States. Alaska is in the Western United States, on the northwest extremity of the country's west coast. A semi-exclave of the United States, it borders the Canadian provinces of British Columbia and Yukon to the east and shares a maritime border with Russia across the Bering Strait (State of Alaska, n.d.). Over half of its population lives in three cities (Anchorage, Fairbanks, and Juneau), and over 86% of Alaskan communities are not connected to the road system. Clearly, the aircraft can provide utility in such a place ("Alaska Mapping," 2017, p. 48).

Aviation is an essential part of life in Alaska. From moving goods, services, and people to providing lifesaving resources and supporting the oil and gas and tourism industries, aviation and the airplane have been critical parts of the State's development from its early days. The following is an example of an operational narrative for an Alaskan operator. While it is not based directly on any one specific operator, pilots who fly in the State will recognize the prevalence of some of its characteristics.

In rural Alaska, numerous examples of large communities are surrounded by smaller villages. These large communities typically average less than 5,000 residents. They have large airports serviced by Part 121 air carrier service, frequently with multiple runways and different approach types. Some have Air Traffic Control facilities or their subset "radio" operators and state maintenance personnel on the facilities.

The towns themselves are social and economic hubs out in the bush. They have more advanced services like fire departments, hospitals, groceries,

 DOI: 10.1201/9781032678863-2

larger schools, churches of different religious sects, and community centers. Surrounding these towns are smaller villages, which rely on the large communities for access to resources outside of the bush.

The aviation model is hub and spoke, and the larger communities are home to pilot bases for aviation operators. Many companies may compete for the same business and have similar operating models and philosophies. They fly to the small villages surrounding the hub and transport cargo and people in their varied fleet types. Many companies offer shift schedules—pilots stay in pilot housing and can work a 2-week on/2-week off schedule. Because of the size of the towns, the pilot groups tend to develop their own shared community—employees on the same shift develop camaraderie among each other due to the nature of living and working with colleagues.

In this example, the operator runs various aircraft types in passenger and cargo operation configurations, namely single-engine pistons. The operator also has a small fleet of single-engine turbine aircraft like the Cessna Caravan, a favorite of bush operators. They fly under Part 135 *regulations and are certified to fly both IFR and VFR. They hire low-time pilots who want to build flight time in the bush before transitioning to the airline industry.*

Turnover among the younger group of pilots may be high. They get their flight time and get out. Alternatively, some pilots choose to make this style of flying their career. Developing relationships between more experienced pilots and the younger generation is critical for maintaining a positive safety culture within the bush flying community in Alaska.

A "day in the life" for a pilot at this company typically involves multiple legs of flying between villages. The pilots are responsible for loading and unloading their cargo and assisting with passenger concerns. They communicate with other pilots and their flight followers for schedule updates and questions regarding operational control or weather.

Many pilots are only eligible to fly during VFR conditions. Long periods of low, cloudy weather can punctuate a shift. When the weather is good enough to fly, there is a frenzy of activity at the airport as planes and pilots try to launch for their flights.

Another example of a common type of operator in the bush supports the tourism industry. Let us review this exemplar operator, who does flightseeing tours in one of the National Parks in Southeast Alaska:

> *Southeast Alaska is known for its fjords, islands, and stunning landscapes. The challenges of flying in this part of the State are augmented by the weather patterns that frequent the area. Southeast is well known to Alaskans as being the rainiest part of the State, and for good reason—it is part of the largest temperate rainforest in the world. The Misty Fjords National Park is in the area and draws a healthy number of visitors yearly, mainly through commercial cruise ship tourism. Operators leverage deals with cruise ships*

to provide passenger air tours—multiple operators in the Southeast have similar business models. Pilots may fly multiple flightseeing tours per day if weather conditions allow.

The primary transportation modalities between Southeast Alaska communities are aircraft or boats. Aircraft on floats accommodate water landings and access to communities without paved airstrips. Many operators fly types like the DHC-2 Beaver or DHC-3 Otter models. These aircraft have been quintessential to operating in Alaska for half a century. Due to their age, maintaining them and finding parts is a challenge that will only get harder as the fleets mature.

The communities in the Southeast, much like those in Western Alaska, are small and close-knit. Being far from the large population centers of Anchorage and Fairbanks has its challenges but also allure for people who want to stay off the beaten path and experience coastal Alaska living.

Southeast Alaska's tour operators have banded together to create voluntary rules for operating near Misty Fjords National Park. These rules are not regulatory but were created to help reduce the number of accidents and incidents between tour operators in the area. Despite these changes, high-profile accidents have occurred in the area since their inception. These accidents will be discussed in a later chapter on Alaska case studies.

FLY ALASKA: HISTORICAL PERSPECTIVES AND OPERATIONS SAFETY

Historical Perspectives

Aviation and Alaska have been synonymous since the state's earliest days. Early pioneers of flying in Alaska were referred to as *bush pilots*; their eagerness and thirst for exploration made them a lasting symbol of Alaskan aviation. They were among the first to step foot in new areas and discover and explore facets of Alaska that had yet to be untouched.

A notable pioneer bush pilot was Harold Gillam, who began air service to the Wrangell Mountain region of Alaska. He was known for his willingness to fly in all weather—even that which most other pilots would choose to avoid. He was fearless in his aviation endeavors and left a legacy still noted in Alaska aviation circles (Ringsmuth, n.d.). Bob Reeves was another pioneering bush pilot in Alaska who revolutionized air travel in the region with his business, Reeve Aleutian Airways. Reeve Aleutian Airways was started in the late 1940s and operated until 2000. Bob Reeves was famously known as "the glacier pilot" due to his impressive number of glacier landings throughout his flying career (Shiner, 2013).

Others, such as Carl Ben Eielson and Russel Hyde Merrill, contributed immensely to the development of bush flying in Alaska. Eielson was the first to fly over the Arctic Ocean and established regular airmail service in Alaska. Eielson Air Force Base, near Fairbanks, Alaska, is named after him due to his achievements (Shiner, 2013). These bush pilots of early Alaskan aviation have become legendary throughout Alaska. Streets, public buildings, airports, and even a local rock band are named after early bush pilot pioneers.

Alaska, nicknamed the Last Frontier to promote tourism after the end of WWII, has long had vivid imagery of remoteness and "edge of the map" type vistas that draw the imagination of people from outside Alaska. This imagery was purposeful; the business benefits were used to help develop Alaska into a tourist destination (Ringsmuth, 2015). Alaska is rich in natural resources, and its geographic location was (and still is to this day) strategic for business and political franchises (Kraus, 2020). Alaska was strategically positioned for military aviation due to its proximity to the USSR and China during the height of the Cold War (Cloe, 2015).

The aircraft was critical. It allowed access to the far reaches of the state and the untapped natural resources limited by Alaska's underdeveloped road system. It was instrumental in kickstarting industries like oil/gas and gold, as many mining sites could only be reached by air (Ringsmuth, 2014). Additionally, Alaska's location made it an essential stopover for trans-Pacific flights during the jet age; Anchorage, Alaska, was perfectly positioned as a strategic refueling point. Most major U.S. airlines currently serve the Ted Stevens Anchorage International Airport (PANC). It is an essential hub for Alaska Airlines, which provides services to many of the larger towns across the state.

However, Anchorage is most recognized as a central air cargo hub, owing to its strategic location on the shortest route between many points in North America and Asia. PANC is one of the busiest commercial cargo hubs globally (CAPA, 2019). Though the initial "gold rush" era of aviation has passed, the relics of aircraft, attitudes, and imagery remain today in modern aviation applications.

Operations Safety

The operational safety record of aviation in Alaska is well-known—and unfortunately, it is subpar. Aviation accidents and incidents regularly make the front page of both local and national media. Studies by the National Institute for Occupational Safety and Health (NIOSH) indicate that between 1990 and 2015, Alaska accounted for 36% of all commuter and air taxi accidents in the United States, with 20% resulting in fatalities (NIOSH, 2018). Comparing data from the NTSB and the Aircraft Owners and Pilots Association (AOPA), the largest G.A. advocate in the United States, Alaska's accident rate is 2.35 times higher than the rest of the United States, with a fatal accident rate of 1.34 times (AOPA, 2021). These are staggering statistics that should not be overlooked.

These studies and analyses pinpoint a few reasons for Alaska's relatively high accident rate: the sheer number of pilots in Alaska and the number of G.A. activities outnumber those in other states. More flying translates to more opportunities for risk, and many of these accidents involve pilots in their private aircraft. Other contributing risk factors include dramatic and rapidly changing weather, hostile terrain, and mission pressures on pilots who provide service in remote areas with minimal infrastructure.

The critical nature of air transport to the Alaskan way of life means that safety incidents and accidents have far-reaching effects on the communities of Alaska. As highlighted in the earlier chapter, CFIT and "VFR into IMC" accidents have been one of the causal elements endemic to aviation accidents in the state (Mode et al., 2012; NTSB, 2017a, 2017b, 2018). In a study on the psychological factors affecting pilots' decisions to fly into deteriorating weather, Madhavan and Lacson (2006) suggest that causes of "VFR into IMC" accidents include various human factors. These include situation assessment, risk perception, and intrinsic and extrinsic factors like social and motivational pressures. Decisions to fly "VFR into IMC" are made more dangerous due to the environment in Alaska, where weather reporting and the IFR infrastructure are limited.

To put this concerning historical trend into perspective, in June 2015, a sightseeing tour aircraft impacted terrain near Ketchikan, Alaska, and killed all passengers on board (NTSB, 2017). The NTSB investigation revealed that inadequate decision-making and organizational culture contributed to the accident (NTSB, 2017). The following year, in October 2016, a separate accident killed all three occupants of an aircraft when it impacted terrain near Togiak, Alaska (NTSB, 2018). The NTSB's final report lists inadequate aeronautical decision-making and inadequate regulatory and organizational oversight as factors in the accident (NTSB, 2018). In August 2021, yet another fatal accident occurred in Southeast Alaska, near Ketchikan when a flightseeing tour impacted terrain in marginal weather conditions. The NTSB again lists inadequate aeronautical decision-making and the FAA's reliance on a voluntary safety program as factors in the accident (NTSB, 2022).

Organizational, Psychosocial, and Cultural Factors That Influence Operational Safety

What could be causing the rather tragic recurrence of aviation accidents in remote areas like Alaska's backcountry? Extant studies suggest organizational, psychosocial, and cultural factors influence decision-making, and invariably safety, among pilots who operate in the bush. For example, plan continuation errors have been suggested to stem from emotional and stress-induced influences. These factors can bias decision-making toward riskier and less rational behavior (Causse et al., 2013). When risky behaviors lead to successful outcomes, biases may develop among pilot groups and become a cultural norm despite company policies advocating against such behaviors (Transport Canada, 2002; Lai et al., 2019).

Sometimes, complex situations with opaque outcomes, time pressures, and influences from cultural norms also create obstacles toward decision-making (Strohschneider, 2002). Chronic psychological stressors can adversely impact decision-making and compromise flight safety (Dismukes et al., 2015). Other factors, which include company and peer-based influence, along with pilot attitudes and motivations, may lead to poor decision-making during flight operations and explain factors in aviation accidents (Causse et al., 2013; Noort et al., 2021; Orasanu, 2010; Strohschneider, 2002; Transport Canada, 2002).

In an earlier qualitative study, Bearman et al. (2009) examined both direct and indirect operational pressures faced by pilots in Alaska. Twenty-eight (28) Part 135 pilots were interviewed, and the study's results suggested that recurrent themes of organizational culture, organizational processes, and resource allocations influenced pilot decision-making. The Bearman et al. (2009) study focused on organizational variables in their initial analysis and did not probe into social and personal influences on decision-making among the study participants. Using the same interview data, Paletz et al. (2009) did a secondary analysis to understand the role of motivations in pilot ADM. The results suggested that personal inconveniences could have a significant influence over ADM.

From a more globalized perspective, Leigh and Rissanen (2010) used qualitative research data from interviews to evaluate the recruiting and training practices of pilots flying in remote parts of Africa. Their findings suggested that most negative motivational influences were related to environmental, behavioral, and cultural factors. Examples of these influences were the excessive difficulties of bush flying, meeting passenger expectations, and pressure from customers and the pilots' organizations to break the regulations. The results further reiterated that pressures to complete flights were due to management practices and, partly, to the pilots' self-motivation.

As a follow-up to these studies, researchers Michalski and Bearman (2014) explored factors affecting pilots' decision-making in Outback Australia using semi-structured interviews with 12 participants. The authors suggest that organizational, social, and personal factors strongly influence pilots' decisions to accept flights. The weak organizational culture, resulting from low safety standards, influenced pilots to make risky decisions. Personal and social influences, like career ambition and pressures from peers and customers, were significant to the pilots' decision-making.

The findings from qualitative studies, which provide more nuanced explanations and individualized perspectives, emphasize that a myriad of different psychological, social, and cultural factors influence Part 135 pilot decision-making. Assessing and understanding how cultural and psychosocial factors influence aeronautical decision-making among Part 135 pilots flying in Alaska can provide valuable insight into optimizing operators' operational decision-making skills.

Process monitoring and ensuring progressive improvement in the safety landscape are crucial. This applies equally to operational safety management in the Alaskan environment. Aviation Safety Management Initiatives can provide data-driven and effective policies that help reduce risks from operational hazards to acceptable limits.

AVIATION SAFETY MANAGEMENT INITIATIVES—EFFECTIVENESS AND LIMITATIONS

There have been numerous initiatives to improve aviation safety in Alaska. In 2000, U.S. Congress started the Alaska Interagency Aviation Safety Initiative program. It was developed to help address the staggering accident rate in Alaska—at the time, it was nearly five times the rate in the rest of the United States (Berman et al., 2005). This program made recommendations that started or enhanced existing programs for aviation safety and used federal funds to meet these objectives. These programs, like the Medallion Foundation and the Capstone program, were prevalent in the state for decades and have left lasting impacts.

The Capstone program was the testbed for the Automatic Dependent Surveillance Broadcast systems (ADS-B). Developed in 1999, this program received $11 million in federal funds (UAA, 2005). ADS-B is a critical component of aviation safety today. It uses ground-based and airborne systems to broadcast aircraft position data in conjunction with onboard aircraft systems. This helps other pilots and air traffic control mitigate traffic conflicts. The program was designed to meet four operational objectives:

1. Reduce CFIT accidents,
2. Reduce traffic conflicts and accidents,
3. Enhance flight weather information,
4. Address shortcomings in the IFR infrastructure (UAA, 2005).

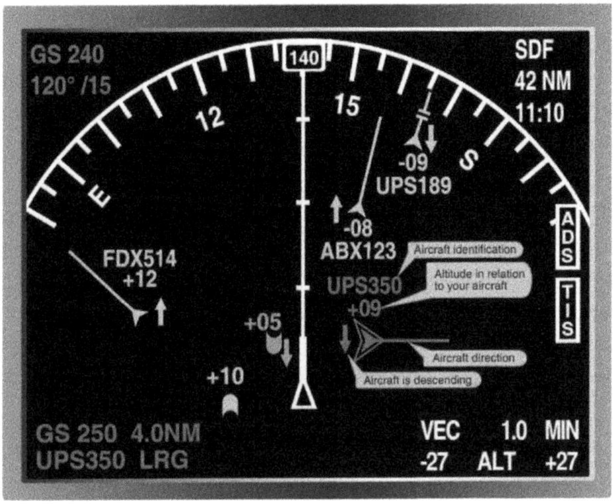

FIGURE 2.1 An ADS-B display showing multiple aircraft targets

Source: FAA, 2012.

For Phase I of the program, equipment capable of ADS-B was installed in aircraft flying in western Alaska, specifically in the Yukon-Kuskokwim Delta. The success of Phase I encouraged program officials to expand into Southeast Alaska in Phase II. This technology is still used today—many operators across the state continue to use the legacy Capstone equipment.

Most operators in Alaska should be familiar with the FAA Aviation Weather Camera Project. This project was started in the late 1990s to provide pilots with real-time weather data. The cameras are installed in strategic areas across the state and can be used for flight-planning purposes. They show live photographs of current conditions and a comparison against prominent visual landmarks in the area (FAA, 2020). The project has expanded outside of Alaska—weather cameras can be found across mountainous and challenging geographic areas such as Hawaii and other areas in the contiguous United States.

Established in 2001 by the Alaska Air Carriers Association (AACA), the Medallion Foundation was a non-profit organization formed to "improve pilot safety awareness" and "lower air carrier insurance rates." The program was defunded in 2019. The foundation's primary goal was to decrease aviation accidents in Alaska by offering safety resources to individuals and organizations, including students and professionals. Their organizational model for aviation safety was called the "Shield Program," which awarded operators for meeting safety objectives as they participated in the program (Mondor, 2021; Hollander & McGee, 2019).

Receiving a Shield award came with benefits—cost savings on insurance was one of them, and marketing opportunities were a second. There were five elements to the program:

1. Safety management system (SMS) and safety program implementation,
2. Training and checking,
3. Operational risk management and risk assessment,
4. Maintenance and ground personnel training,
5. Internal audit system (Alaska Aviation, 2005).

These safety programs included training on flying in mountainous terrain, CFIT awareness training, and risk assessment matrix development. Participation in these programs was voluntary, and operators were provided free resources to aid implementation.

The Capstone Program has reached its termination point, and the Medallion Foundation has closed its doors. The benefit of programs like these has been monumental in providing operators with materials and technologies to help improve aviation safety in the state. The NTSB has critiqued FAA oversight of the Medallion Foundation as a contributory cause of accidents. Countering this, proponents say that the Medallion Foundation had no regulatory powers and was a completely voluntary program. Despite these challenges and the program's closure, proponents in the state enumerate the benefits of the Medallion Foundation (FAA, 2021b).

The NTSB has kept a close eye on trends in aviation safety in Alaska. This was recently the subject of an NTSB safety roundtable (2019a). As mentioned earlier, Part 135 safety was on the NTSB's most-wanted list for 2019–2020 (NTSB, 2019b). The Most Wanted List for 2021–2022 also addressed Part 135 safety and highlighted the need for SMS to become a required component of safety programs for revenue passenger-carrying operations (NTSB, 2021).

The FAA Alaska Aviation Safety Initiative, the latest study on aviation safety in Alaska, aimed to gather stakeholder input and develop projects to enhance aviation safety in the state. The report was a multi-year, interdisciplinary study in response to the NTSB roundtable report from the previous years. The study interviewed essential Part 91 and Part 135 stakeholders in Alaska to elicit their concerns and the challenges they faced as operators. The four key areas addressed in the study are as follows:

1. Environment: weather, terrain, and lack of reliable weather reporting stations,

2. Aircraft fleet: aging technology and aircraft, costs associated with upgrades to fleet types or avionics,
3. Infrastructure: communications, navigation, and surveillance,
4. Operations safety management: pilot experience and training (FAA, 2021b).

This chapter discussed the unique landscape of Alaska. Aviation is an essential component of the state for personal enjoyment, economic development, and transportation where there is no other infrastructure. While aviation has become part of life, so too have the significant safety challenges that continue to impact operations. Addressing these challenges is paramount to ensure the continued success and safety of the people who rely on aviation in Alaska.

3 Theoretical Perspectives of Human Factors That Influence Aeronautical Decision-Making in Remote Operational Settings

A well-known joke goes like this: How do you know a pilot is in the room? They will tell you. A common understanding of pilot personality is that pilots have big egos and the "right stuff" to get the job done. This might stem from early aviation pioneers who had to overcome difficult odds flying old equipment in challenging environments with minimal training.

It is essential to look at the theory behind why certain factors influence ADM in remote operational settings—we will explore these factors comprehensively in this section. Most of the theoretical perspectives discussed apply to pilots in all operational environments. However, given the extreme environments in which remote pilots fly, it is crucial to identify how these factors impact ADM and, ultimately, their safety behaviors. We will examine how

DOI: 10.1201/9781032678863-3

smaller working groups, fewer organizational constraints, and more diverse environments and pilots play a role in pilot decision-making.

HAZARDOUS ATTITUDES AND PILOT PERSONALITY

So, what is the difference between attitude and personality, and how are they related? Personality refers to the intrinsic character traits that make up a person. It is their enduring qualities and demeanor, the "essence" of who they are. Personality influences a person's thoughts, motivations, and behaviors as they experience different situations—it is a dynamic and organized set of traits (Ryckman, 2013).

Personality has been suggested to remain relatively unchanged due to its development during a person's upbringing and childhood (Gao & Kong, 2016; Helmreich, 1984; Ryckman, 2013). Our personality influences our relationships with family members, friends, and colleagues. Personalities affect how we think, feel, and invariably provide a behavioral metric for our colleagues regarding what to expect as they interact with us (SocialSci.Libretext.org, 2021).

Personality plays a critical role in many aspects of aviation operations, and various personality characteristics have been identified as crucial for pilot training and aviator job success (Barron et al., 2016; Breuer et al., 2023; King et al., 2012). People possessing conscientiousness, achievement-striving, and assertiveness may be drawn to the piloting profession. However, these traits do not necessarily determine pilot ability (Fitzgibbons et al., 2000; King et al., 2012; Sutin et al., 2009).

For example, let us suppose a pilot is known for always running a comprehensive pre-flight checklist before taxiing out for departure. Suddenly, they start skipping the pre-flight checklist altogether. This directly conflicts with their personality and established behaviors—they are usually meticulous and conscientious. As a result, other pilots who know this person may perceive something odd or wrong with the behavior. On the other hand, if the pilot was known to be slack in pre-flight checklist discipline, this observation might not be as surprising or noteworthy.

Likewise, suppose an ordinarily even-tempered captain reprimands you for something minor on the flight deck. In that case, you may believe there is something more to his or her anger since this is not a typical personality trait.

Attitude is our outward demeanor, influenced by positive and negative reinforcements from the environment around us (Hunter, 2005). It is how we respond to stimuli and interact with people and situations. Our attitudes are generally favorable or unfavorable opinions toward people, things, or situations and are affected by the environment we were brought up in and by our individual experiences. For example, some pilots may have negative attitudes toward bush flying, but their attitudes may change as they gain more experience flying in the bush environment. If the pilot has a bad experience, they may develop a negative attitude around bush flying.

Attitude better indicates pilot successes in training and development than personality. Attitude influences our behaviors and decision-making, and when appropriately adjusted to the environment, it can be an excellent hazard barrier. Hazardous attitudes are discussed in pilot training from the first few flight lessons as part of any private pilot curriculum (Neff, 2022). The FAA has identified five of them as deleterious to pilot performance: antiauthority, impulsiveness, invulnerability, macho, and resignation (FAA, 2023). Hazardous attitudes contribute to poor pilot decision-making.

The recognition of hazardous attitudes is the first step toward neutralizing them. After recognizing a thought as hazardous, the pilot should label it as such and then state the corresponding action to avoid it. Through thoughtful training, pilots should learn to recognize, recover, and avoid hazardous attitudes. This is typically a fundamental element of early pilot training, but deeper conversations on how pilots make decisions are essential throughout a pilot's career. Let us look briefly at each hazardous attitude and see how they may impact aviation safety in a remote setting:

Anti-authority

Anti-authority is the willingness to disregard rules and regulations. It is a pilot's superiority over or above the rules and willingness to break policy or procedure to accomplish a course of action they feel is the best route. This behavior may lead to a few things: one, it may lead to more violations among pilots (English & Branaghan, 2012), and two, it may cause the erosion of effective flight deck communication.

You may ask, *but why is this important in a single-pilot environment, like many remote operations?* Authority extends beyond just the inter-flight deck dynamic. More obviously, authority can refer to the governing bodies and regulations promulgated to make our aviation infrastructure efficient and safe, as well as the operational procedures and policies developed within an organization's structure. Procedures are written to mitigate the associated risks in an environment with particular and niche operational challenges.

In the bush, external pressures may influence pilots to make substandard decisions—customer or company pressure to complete flights may lead to decisions that violate operational procedures or federal regulations. There has been discussion in the bush pilot circles that often, the safer choice means the not-legal choice. This certainly deserves more recognition on the part of regulators to address. Is a lack of infrastructure contributing to these decisions? Is it an improper policy? Where are the gaps that pilots must navigate around, and why are they being asked to make decisions in such grey areas?

Compared to the rest of the United States, antiauthority and invulnerability have been components of violations nine times more likely to end in fatal accidents in Alaska (Detwiler et al., 2006, 2008). When considering continued VFR into IMC with CFIT as an outcome, antiauthority and invulnerability are especially lethal, accounting for 80% of fatal accidents in Alaska (NTSB, 2020).

Invulnerability

The invulnerable pilot feels immune to accidents, risks, or adverse outcomes from safety events (Adhikari, 2015). This trait leads to a lax attitude toward safety; pilots who display it often brush off risk and consequences. These pilots may accept flights and launch into challenging weather or conditions regardless of the risk it may pose to them, their crew, or their passengers. A pilot risk perception and tolerance study suggested that pilots who do not recognize risks associated with adverse weather are more likely to engage in risky decision-making (Hunter, 2002).

Another study suggested that pilots who have successfully flown VFR into IMC are likelier to repeat the behavior (Wiggins et al., 2012). Finally, a study on airline pilots that discussed personality and risk tolerance suggested that pilots with more experience are likelier to adopt a higher risk tolerance. They are more willing to accept flights in challenging and potentially lethal weather conditions, therefore embodying the attitude of invulnerability (Makarowski et al., 2016).

These three studies suggest that invulnerability goes hand in hand with experience. However, it is essential to note that a risk appetite is part of an attitude that may be trained out of pilots or mitigated with proper procedures and policies. Furthermore, strict adherence to these policies at all levels of an organization—from line pilot to CEO—helps mitigate the invulnerable attitudes that may accompany highly experienced individuals.

Macho

Social science literature describes *machismo* as a social construct in which a person tries to prove manhood and self-sufficiency (Stewart, 2006). In the aviation world, this translates to pilots trying to prove themselves. It can manifest at all levels of aviation, including training, career, and recreation.

In the bush, a social hierarchy exists. Pilots with more experience are frequently paired with newly-hired pilots and help aid their development—this could give the more experienced pilots an illusion of control over the newly hired pilots. New, young, or inexperienced pilots may work with (or fly with directly) pilots who display these macho attitudes and, as a result, may not be willing to speak up or voice safety concerns. This creates barriers to communication and crew resource management in the flight deck. These green pilots could end up complying with risky behavior due to the influence of seniority in the flight deck (Michalski & Bearman, 2014; Noort et al., 2021). Overconfidence tends to go hand in hand with macho; pilots may attempt flights purely from the admiration they may receive from their peers (Murray, 1999).

Macho pilots sometimes let ego get in the way of good decision-making. There is often a fine line between overconfidence and being confident in making the right decision—when a macho attitude influences decision-making, the rationale is typically not indicative of healthy safety-risk decision-making.

Impulsiveness

Impulsive behaviors are when we act before thinking or choosing the first action that may come to mind. In the aviation world, impulsiveness is most frequently associated with antiauthority. Clarke and Robertson (2005) discussed several personality traits associated with accident involvement:

1. Carelessness and lack of self-control,
2. Impulsiveness,
3. Lack of respect for authority.

These three traits encompass the broader hazardous attitudes frequently occurring in aviation accidents and incidents, leading to the most adverse safety events.

Resignation

Resignation does not get much attention in the aviation literature on decision-making. This may be because resignation is the *lack* of action by a pilot. It is their hopelessness, giving up on a situation, or accepting that fate has more control over their circumstance than they do. This attitude can be dangerous in

The Five FAA Hazardous Attitudes and Antidotes		
Type of Attitude	Description	Antidote
Antiauthority	"Don't tell me..." *Pilots with an anti-authority attitude tend to believe that rules, regulations, and safety procedures don't apply to them. For example, an anti-authority pilot may neglect their checklists or refuse to take instructor advice. Be advised that having an anti-authority attitude is different from simply questioning authority. Pilots always have the prerogative to speak up to authority if they believe a mistake has been made.*	Follow the rules; they're usually right.
Impulsivity	"Do something quickly!" *Pilots who 'feel the need to do something, anything, immediately.' Instead of taking a moment to think things through or select the best alternative, a pilot with an impulsive attitude does the first thing that comes to mind. Reacting too quickly can lead to irrational decisions, such as skipping a preflight or rushing to get home (or off the hill) despite inclement weather. Pilots are encouraged to take time to evaluate their options before choosing a course of action.*	Not so fast- Think first!
Invulnerability	"It won't happen to me...." *Many people—not just pilots—fall into a pattern of thinking that accidents happen to others but never to them. This attitude of invulnerability can become a safety concern when pilots fail to consider the risks of their actions.*	It could happen to me!
Macho	"I can do it." *Pilots with a macho attitude are always trying to impress others and prove themselves by taking unnecessary risks. Both men and women are susceptible to this attitude, which leads to foolish and often dangerous behavior. While pilots must have high confidence in their abilities, it is essential to avoid becoming overconfident and adopting a macho attitude.*	Taking chances is foolish.
Resignation	*"What's the use?"* *Pilots with an attitude of resignation lack the confidence and conviction to believe they can make a difference in what happens to them. These pilots tend to give up easily when faced with challenges and don't take criticism well. Bad things are attributed to luck. They think "someone is out to get them when things don't go well." This attitude is particularly dangerous for pilots in an emergency because they may believe they are helpless and resign to their fate instead of taking action. Such pilots can go along with unreasonable requests just to be the "nice guy."*	I'm not helpless. I can make a difference.

FIGURE 3.1 The FAA's five hazardous attitudes

Source: Reproduced from FAA, 2023, pp. 2–4.

an emergency because a pilot may believe they are helpless and resign to their fate instead of taking action. They may also not be assertive or speak up in a multi-crew environment to proffer their perspectives during an emergency or procedural non-compliance by another crew member. Frequently, the hazardous attitude of resignation is precipitated by physical or mental stressors.

IMPACT OF EGO AND OTHER ATTITUDINAL FACTORS ON PILOT DECISION-MAKING IN THE FLIGHT DECK

Often, the punchline to jokes or the subject of aviation-related pop-culture media, pilots and ego are synonymous. Ego tends to become a factor in occupations with high-risk, high-reward job objectives. Other fields like medicine or elite levels of athletics may see an element of ego among their significant players. Ego makes an individual passionate and ambitious, but letting it go too far can become obnoxious and even dangerous (O'Bryan, 2011).

A world-class aerobatic pilot, Patty Wagstaff, states,

> I know my ego is tied into my flying more than in any other area of my life because it is part of my identity and what I'm most proud of. I think I am a sensible, cautious, professional pilot. Most pilots I know are. But, in the back of my mind, I always have to check to see who's driving the bus—me or my ego—so I don't do anything dumb (again)! I know I've said it before, but there's just not enough room in the cockpit for both me and my ego.
>
> (Wagstaff, 2016)

Pilots' tendencies to be assertive and achievement-striving (Fitzgibbons et al., 2000) are personality traits associated with ego. Imagine having a career where your performance is evaluated constantly—your company evaluates your skills yearly in a test designed to challenge the limits of your decision-making. Keeping (or losing) your job may be the outcome.

On the day-to-day, your performance is evaluated by your partner in the flight deck. Quality debriefing among the crew plays a crucial role in the skill development of both new and experienced pilots. Furthermore, if you have a bad landing, you can be relatively assured that one of your passengers will let you know! Ego, therefore, is something that pilots feel must be protected.

An attempt to protect ego on the flight deck by pilots can adversely affect crew coordination and communication. A captain's ego can prevent open discussion with his crew—admitting mistakes could show a lack of knowledge or weaknesses in his abilities. A first officer's ego could prevent him from speaking up about possible errors for fear of disrespecting or embarrassing his captain. A cockpit with large egos can have disastrous consequences without open communication (O'Bryan, 2011).

There is a concept called *Loss of Face*, which is essentially protecting ego and self-image. An individual's assessment of how others view him is "*face*"

(Murray, 1999). Protecting "*face*" is a natural part of human nature. For the pilot, it may be augmented due to the self-correcting nature of the job and how pilots are continually critiqued and evaluated. An Alaskan study on pilot decision-making showed that pilots had an aversion to social disapproval; this aversion was less subtle in situations where a pilot would look bad or admit to mistakes or defeat in front of their peers (Paletz et al., 2009). Other studies show pilots who are unwilling to ask for help even when they know they are lost because they do not want to "broadcast their failure as pilots" (Gilbey & Hill, 2012, p. 785).

If protecting your ego and "face" is so important, it follows the reason that self-incriminating information could be withheld in safety-related settings. These could be failures, poor decisions, or mismanaged situations in which a pilot could feel embarrassment by disclosing their actions. Crew resource management was designed to eliminate barriers during flight crew communication.

Withholding critical safety information could lead to a breakdown in communication between pilots; it is ultimately a dangerous aspect of how ego impacts safety in the flight deck (O'Bryan, 2011). Even in a single-pilot operation, sharing safety-critical information (including errors!) is the best practice for improving safety across an organization and enabling other pilots to make the best decisions for their flights and operations.

Pilots are more likely to make high-risk decisions when they have gotten away with making similar decisions in the past. This creates an unrealistic optimism toward risk and inflates the pilot's sense of ego and narcissism—if they have done it once, they can do it again. The greater willingness to take risks leads to adverse safety outcomes.

GOAL SEDUCTION, PLAN CONTINUATION ERRORS, AND PILOT DECISION-MAKING

Goal Seduction is not as sexy as it sounds. It is how a desired outcome can persuade decision-making. Suppose the target outcome or goal has too much of an influence on decision-making. In that case, it may lead to errors during flight operations and an incident or adverse safety event (Bearman et al., 2009).

Goal seduction can pressure decisions at the expense of safety and change priorities in favor of a specific outcome. These decisions can be made to reach operational objectives and productivity goals, outperform the competition, fit in socially, protect one's ego, or make a paycheck—all common themes throughout remote aviation settings. Goal seduction in the Alaskan context was identified when pilots had to rescue others, meet significant others, and handle time-related constraints and financial pressures (Bearman et al., 2009; Maglio et al., 2016).

Plan Continuation Errors

Plan continuation error is the tendency for a person to continue with a plan despite changing conditions, which should dissuade them from proceeding (Orasanu et al., 2002; Causse et al., 2013). The perceived negative consequences of switching away from the original goal influence the continuation of the original trajectory. Dekker (2014) suggests that plan continuation errors occur when cues used to formulate the initial plan are considered very strong.

For example, if the plan seems solid based on current information, later indications may not be viewed equally in decision-making. In aviation, there are numerous behavioral traps associated with plan continuation errors. "Get-there-itis," VFR into IMC, and neglecting to initiate a go-around all relate to plan continuation errors. These three examples place pilots in a position where adverse safety outcomes could potentially come into play (Harris et al., 2022; Velazquez, 2018).

Numerous aircraft accidents in remote settings have partly been attributed to plan continuation errors. VFR into IMC scenarios are particularly prevalent. On February 24, 2016, a Tara Air De-Havilland of Canada DHC-6 turboprop aircraft continued toward its destination after encountering adverse weather. The crew was on a VFR daytime-scheduled domestic passenger flight from Pokhara to Jomsom, Nepal. They repeatedly entered clouds while trying to reach their destination, ultimately impacting rocky terrain at an altitude of almost 11,000 feet just 15 minutes after takeoff.

The aircraft was destroyed, and all on board were killed. The accident investigation attributed this accident to the crew's repeated decision to fly in the clouds and their deviation from the intended route after losing situational awareness. The report also concluded that the crew became spatially disorientated and failed to respond to repeated Enhanced-Ground Proximity Warning System (EGPWS) cautions and warnings about the impending high terrain (Aircraft Accident Investigation Commission, Government of Nepal, 2016).

Foot-in-the-Door Persuasion

The foot-in-the-door technique is a compliance tactic suggesting that agreeing to a small request increases the likelihood of agreeing to a second, more significant one (Freedman & Fraser, 1966; Pascual et al., 2013). Once you start something, it is significantly easier to keep momentum going. When asked to do something small, the door opens for more extensive and significant requests, and a person is more likely to agree to more (Dolinski, 2012). Pilots are more likely to take flights in conditions they know they have successfully flown in previously or that their peers have successfully completed. Management and other pilots may take advantage of this and exploit this social theory to achieve operational goals.

SOCIAL THEORIES RELATED TO OPERATOR DECISION-MAKING

Social cognitive theory is the theory that analyzes the interactions between three pressures on decision-making: behavioral, environmental, and personal (Bandura, 1986; Stajkovic & Sergent, 2019). The theory conceptualizes a belief system that guides behavior toward desirable pursuits and away from undesirable consequences. As people set goals, devise accordant courses of action, and anticipate outcomes, they act on this thread of beliefs. The theory suggests that as we operate in the environment, we monitor and analyze our actions, reflect on consequences, and react to correct our course toward a desired outcome.

These behaviors rely on self-awareness, self-reflection, and adaptive self-regulation (Bandura, 1986). The theory emphasizes observed behavior—how do our actions change based on the direct experiences we observe daily? The more we relate to the subjects, people, or environment we are surrounded by, the more likely we will be influenced by their behaviors. An encouraging note: positive outcomes are more likely to influence behavior than adverse outcomes would deter it (Wood & Bandura, 1989).

Since moving to Alaska years ago, I (Dana) have been surrounded by outdoorsy people. As someone who grew up in the Midwest, I was not exposed to the mountain running, hiking, backpacking, and skiing that I do on my days off now. In this case, my environment is a massive primer for the type of life I now lead. My interpersonal relationships with like-minded people augment these lifestyle decisions, and on a personal note, I feel good when I spend time outside. The skills I have gained in the backcountry were largely from observing others around me. This is an example of social cognitive theory as it applies to my personal life.

Social Identity Theory, Collective Self-efficacy, and Aviation Group Behaviors

Social identity theory seeks to explain the cognitive processes and social conditions underlying intergroup behaviors, especially those related to prejudice, bias, and discrimination (Tajfel & Turner, 1979). It further explains an individual's role in terms of membership in larger groups. The groups could be about family, education status, favored sports teams, or anything else. Belonging to a group improves self-image, influences behavior, and begins to define who someone is as a person (Brown, 2000).

Think of this as the "in-group." It is what you identify with, and strong positive connotations are associated with it. The "in-group" is what the theory asserts: we are naturally inclined to perceive in a positive light. Conversely, we are neutral or even hostile toward out-groups. Belonging, therefore, enhances our self-image (Li & Zhao, 2012). The "out-group" is outside our social domain. Not belonging to the in-group has a negative bias—behavior may shift to meet in-group objectives (Flippen et al., 1996).

Belonging to the in-group means a person has specific characteristics (Burford, 2012). They either already meet the objectives of the other people in the group, or they alter their behavior and adapt to the norms—this helps them conform to the group identity (Korte, 2007). Because of the high value of belonging, of "fitting in," individuals may go against their beliefs and behavioral inclinations to belong. Decision-making, therefore, is influenced by group behavior. Similarly, teachings from the group behavior may influence a person more than a training environment or policy could (Schunk & DiBenedetto, 2020).

Another important consideration is the relationship between social cognitive theory, group self-efficacy, and the impact on group behaviors. Within social cognitive theory, collective efficacy determines the ability of a group to reach objectives. This is determined by what groups choose to do, their effort, and how long they can sustain that effort while facing challenging circumstances. It is not important how skilled individual group members may be. If they do not believe they can achieve results as a group, they are less likely to try as hard, more likely to give up, and, at the end of the day, perform poorly as a team (Stajkovic & Luthans, 2002).

Many complex inter-relationships exist among the aviation industry's various occupational groups and sub-groupings. There have been suggestions that collective efficacy is a critically important motivational construct in predicting group effectiveness(Elms et al., 2023). The group dynamics between pilots, flight attendants, management, and maintenance personnel differ. All must interact seamlessly to safely complete flight operations (Ford et al., 2013).

With cohesion between these workgroups, safe operations would be easier to achieve. The emphasis on belonging to an organization instead of a subgroup like "pilot" or "flight attendant" can help mitigate communication barriers between different groups. With a shared objective (e.g., reaching a destination without complications or having to fill out any paperwork), these barriers are even more effectively mitigated (Ford et al., 2013).

Case Studies on Alaskan Aviation Operational Settings

4

INTRODUCTION TO ALASKA CASE STUDIES

Given time and space, we could author an entire Alaskan aviation case studies book. The evolution of flying in Alaska has contributed to a colorful—and often bizarre—case history. At the beginning of aviation in the state, flying was a challenging endeavor that required thinking outside the box. Aviation pioneers had to be highly innovative and ingenious to deal with the complexity of flying in the uncharted regions of Alaska, especially when most of the aircraft had rudimentary technology.

The early aviators were inventive, skilled, and somehow, lucky more times than not. Their attitudes toward flying and the technology available required them to rely on their stick and rudder skills. They were primarily single-pilot in the beginning, and in that period of aviation development, policy, procedure, and safety systems were not yet prevalent in the industry. Much of this attitude has remained today; general aviation pilots are keen to explore backcountry places and push the limits of their more modern aircraft and technologies.

DOI: 10.1201/9781032678863-4

Alaska offers many potential commercial pilots a proving ground to "cut their teeth" and hone aeronautical skills. These pilots also accrue flight time and experience in this beautiful, unique environment. Unsurprisingly, some of these pilots get "sucked in" and choose Alaskan "bush flying" as a life-long career.

If we gave early Alaskan aviation pioneers a crystal ball, we could only imagine their awe at the state of the industry today—they had no safety regulations, minimal practical experience flying, and basic instrumentation in their aircraft. We have come a long way. Improved infrastructure, careful regulation of aviation activities, a history of operational experience, and technological improvements have contributed to significant gains in safety (FAA, 2021). Since the late 1990s, the aviation industry has shifted its paradigm to a more safety-oriented mentality by introducing systems-based safety, proactive philosophies toward aviation operations, and advanced technologies (Transport Canada, 2019).

This genesis of aviation technology and safety systems should reduce the rate of aviation accidents, right? By and large, it has. The first decade of the jet age saw dramatic improvements in fatal accident rates. Since then, safety advancements across the industry have helped continue the downward trend. The 2023 Boeing statistical summary of commercial aircraft accidents reflects this significant reduction in accidents and incidents.

The report suggests a 20-year downward trend (Boeing, 2023). This statistic applies to commercial aviation at large and suggests it is the safest mode of transport. Despite this seemingly impressive data, we will examine three accidents that highlight lingering problems in Alaskan aviation and how their significant factors still impact flying in remote operations.

THE ENDEMIC OF CONTROLLED FLIGHT INTO TERRAIN

CFIT Case Study: Togiak

On 2 October 2016, two pilots began their workday in Western Alaska. They were flying a Cessna 208B Grand Caravan, a popular aircraft model used in the bush. The pilots flew for Hageland Aviation Services, which operated under the name Ravn Connect. Hageland was a Part 135 operator transporting people and goods from their larger pilot bases to smaller villages across Western Alaska in a hub-and-spoke operation.

The operation appealed to pilots at the beginning of their career—Hageland hired low-time pilots who could expect to accrue flight time flying in the bush rapidly. The Second in Command (SIC) had recently finished training and was on his second rotation with the company. Other people drawn to the job were those who enjoyed a shift-style schedule. Pilots typically worked a two-week-on, two-week-off rotation, which provided a particular lifestyle to their pilot employees.

The crew was expected to fly five legs that day, all dispatched under Day VFR regulations. Per the company's Operation Specifications (OpSpecs), only one pilot was required to operate Day VFR. Adding a second pilot allowed the crew to increase their flight time limit from eight to ten hours. Adding a SIC provided an additional person to help load and unload cargo from the aircraft—the company operations procedures expected pilots to work together as a crew when flying in a multi-crew environment.

The crew started their day at their base in Bethel, Alaska. They flew from Bethel to a small village called Togiak, where they offloaded their cargo. Shortly after, they departed on their second leg to a village called Quinhagak. So far, their flights were uneventful. They flew leg two at approximately 4,500 *ft* mean sea level (MSL), which provided ample terrain and obstacle clearance along their flight route. Shortly after landing in Quinhagak and once again offloading cargo, the pilots boarded a single passenger and departed back to Togiak as Flight 3153 along the same route they had just flown.

This time, however, they elected to fly at approximately 1,000 feet MSL, which gave the crew between 500 *and* 700 feet of obstacle and terrain clearance. A mountain ridge approximately 10 nm N.W. of Togiak intersected the direct route between Togiak and Quinhagak. Its highest elevation was approximately 2,485 feet. It was this mountain ridge that became obscured by weather, and this mountain ridge that the aircraft eventually impacted during flight, killing all three occupants onboard.

Perhaps the most intriguing element about this accident is that a second crew of two pilots had a similar flight plan to Flight 3153. They departed Quinhagak for Togiak approximately two minutes after the accident flight. These pilots elected to alter their route slightly to the south to avoid the valley fog that was unexpectedly obscuring the mountain ridge. They were able to stay over lower terrain, out of the weather, and land in Togiak uneventfully.

After landing, the second crew noticed that the first aircraft had not yet arrived despite a slightly earlier departure time. They had not heard or seen Flight 3153 during their flight over the mountains. It was not until the Air Force Rescue Coordination Center received notice of an emergency distress signal that Hageland, the operator, was aware of the possibility that one of their aircraft had gone down.

Commercial and general aviation aircraft in the United States are legally mandated to have an emergency locator transmitter (ELT) on board. The mandate resulted from the 1972 loss of U.S. Representative Hale Boggs and Nick Begich in Alaska after their aircraft crashed and was never found (AOPA, 2018). In the event of an aircraft accident, these devices are designed to transmit a distress signal on 121.5 *and* 243.0 *MHz* frequencies and, for newer ELTs, on 406 *MHz*.

Upon receiving notification of the ELT signal, the company pulled satellite tracking data from the aircraft and realized it had not been updated in over 20 minutes. They re-dispatched the second aircraft to search for Flight 3153 but could not locate it due to cloud cover. Later, after the weather delay, the Alaska State Troopers found the wreckage using one of their helicopters. The Caravan had struck the ridge at approximately 2,300 ft MSL in an extreme nose-up attitude. It was broken into two major sections—the fuselage's final resting place was about 1,500 ft MSL. The right-wing settled approximately 200 ft below it.

The NTSB findings as to the cause (s) of this accident were:

> The flight crew's decision to continue the visual flight rules flight into deteriorating visibility and their failure to perform an immediate escape maneuver after entry into instrument meteorological conditions, which resulted in controlled flight into terrain (CFIT). Contributing to the accident were (1) Hageland's allowance of routine use of the terrain inhibit switch for inhibiting the terrain awareness and warning system alerts and inadequate guidance for uninhibiting the alerts, which reduced the margin of safety, particularly in deteriorating visibility; (2) Hageland's inadequate CRM training; (3) the Federal Aviation Administration's failure to ensure that Hageland's approved CRM training contained all the required elements of Title 14 Code of Federal Regulations 135.330; and (4) Hageland's CFIT-avoidance ground training, which was not tailored to the company's operations and did not address current CFIT-avoidance technologies.
>
> (NTSB, 2018, p. xi)

Analysis

The investigation into this accident was thorough. Hageland's accident history, coupled with the severity of this accident, prompted the NTSB to conduct a one-day investigative hearing on the operation. The hearing explored Hageland's policies for operational control, pilot training, and safety management relevant to the Alaskan operating environment (NTSB, 2017b). In addition, many of the findings from the hearing were used in the final NTSB report for Flight 3153.

Between December 2012 and April 2014, Hageland had five accidents and one runway excursion. Three of these accidents involved operating in IMC while under VFR, commonly a precursor to CFIT. Before Flight 3153, and following the numerous accidents, the FAA and Hageland's principal operations inspector had sent multiple safety requests to Hageland. After an accident on May 4, 2013 (the probable cause for this accident was a continued VFR flight into IMC), the requests ranged from letters asking flights not to be released in adverse weather to a letter questioning the ability of persons with operational control to oversee such a large organization.

There were also questions on whether flight crew training was appropriate for the environment in which the pilots operated. The training program was later suspended following another accident in November 2013. The following year, in April 2014, after the second fatal accident in six months, the NTSB issued an urgent safety recommendation, suggesting that "the recommended action required immediate action to avoid an imminent loss from a similar accident" (NTSB, 2018a, p. 32).

The NTSB highlighted some other factors in the report that contributed to the accident of Flight 3153. First was the routinely inappropriate use of the TAWS inhibit switch and training on its use. Two systems were Installed in the Caravan to alert pilots to potential terrain conflicts. One was a ground proximity warning system (GPWS), and the other was a TAWS. These two systems were independent, providing valuable data to the pilots for enhanced situation awareness. Additionally, there was the option to display a map on the multifunction display on the aircraft panel. The map would show different terrain features and visually highlight threatening terrain when the aircraft was closing in rapidly on such terrain.

The TAWS system had its own internal GPS and terrain database. TAWS is a look-ahead system that alerts the crew to potential threats. This system had an en-route terrain clearance of 700 AGL. Below this, the system would alert the pilot unless the aircraft was in the immediate surrounding area of an airport. Caution alerts are aural and visual and begin one minute from impact with terrain or an obstacle with the synthetic voice alert ("CAUTION TERRAIN"). Warning messages (aural and visual) begin 30 seconds from impact ("TERRAIN, TERRAIN, PULL UP").

On Flight 3153, the crew flew at an altitude between 500' *and* 700' AGL, well within the cautionary range of the TAWS (NTSB, 2018a). The TAWS system included a separate display at the top of the dash. The display had red and amber visual alerts and an inhibit switch, which inhibited both the visual and aural alerts. The inhibit switch stayed engaged until de-selected by the pilot. White text "TERR INHB" was backlit to display the switch's selection (NTSB, 2018).

The TAWS terrain inhibit switch was used, which was customary for flights at Hageland. Their manuals allowed flight as low as 500 *ft* AGL during visual conditions (well below the 700 ft range for caution alerts). At this altitude, nuisance alerts from the TAWS system could distract pilots. Though flights at these altitudes were allowed (and habitual below 1000 *ft*, according to the information gathered from the investigative hearing), the manuals did not instruct pilots on the proper use of the TAWS inhibit switch.

During the post-accident debrief, the chief pilot stated that pilots were authorized to inhibit the TAWS system while flying in visual meteorological conditions where they could see terrain. Given these parameters, pilots would frequently inhibit the TAWS system to reduce nuisance alerts during flight. There was no manual information, company procedures, or checklists to trap this potential threat.

It is believed that TAWS was inhibited on Flight 3153, which would have prevented the system from alerting the crew to the mountain ridge along their flight path. The second aircraft flying that route had also inhibited their TAWS system. The SIC on that aircraft could not recall when they chose to uninhibit the switch to return it to the normal operating position.

Nuisance alerts can cause many problems in operational settings. Managing false alerts impacts a pilot's trust in TAWS, and therefore, the safety element of the system is lessened (Ruskin et al., 2021). When alerts are invalid, they can be distracting, leading to desensitization, and provoke inhibition of the systems and failure to re-activate the system when required (Friedman-Berg et al., 2008).

Due to the prevalence of the TAWS inhibition problem in Part 135 operations, the FAA has issued information to operators who fly under visual flight rules in mountainous terrain. The FAA states,

> Inhibiting warning systems and ignoring warnings, combined with deteriorating weather conditions leading to loss of visual surface reference and situational awareness, has been the cause of some CFIT accidents. In some situations, aircraft impacted terrain that might have been avoided had the TAWS alert feature been uninhibited
>
> (FAA, 2023)

The second and third causal factors were Hageland's inadequate CRM training and the FAA's failure to oversee the CRM training program. Lastly, Hageland's CFIT training program was not designed using current information and did not address current technologies—the program was not tailored to the company's operations overall.

Training is a critical element of pilot skill development. Since the introduction of CRM in the late 1970s and its formal adoption by airlines in the 1990s, training on operating together as a crew has been instrumental in decreasing aviation accidents in the industry (Alexander, 2016; Helmreich &

Foushee, 2019). Hageland had CRM training materials for their pilots, which were delivered during initial and recurrent training in the form of computer-based training modules.

The company CRM training outline included topics like improving interpersonal communication and utilizing resources as a crew to achieve better decision-making. The training included definitions of what was required as the PIC (general leadership materials) and for the SIC (follower duties, which include supporting, monitoring, and respecting the PIC). There was no documented training for the roles of pilot flying versus pilot monitoring or how to work together as a team using this division of duties.

So, while providing CRM training to crews is a crucial element of the training footprint, there were significant gaps in the materials and implementation of these concepts in the flight deck. Farago et al. (2019) suggest that the assessment of some CRM training programs among air service providers is inadequate. They advocated for better training design, delivery, and evaluation for CRM. Farago et al. (2019) also recommended developing assessment strategies applicable to various organizational contexts using scientifically sound metrics.

Many of the flying in the Alaskan "bush" take place in single-pilot aircraft. Pilots enjoy the added freedom they can have as the aircraft's sole occupant and the decision-making freedom it gives them. Adding a second crewmember adds complexity to CRM concepts but heightens safety and situation awareness if the crews can utilize CRM principles effectively. Because SICs were not required crew members on most flights, it is possible that behaving as a single pilot instead of a two-pilot team was naturally ingrained into the mindset and habit pattern of captains at Hageland. While two-pilot crewing may not be the most popular staffing model among pilots, the heightened safety element is a net positive from an operational perspective.

In the United States, Part 135 fixed-wing operators do not have to—and are not required to have—a CFIT avoidance training program. CFIT avoidance training programs are required for Part 135 helicopter operations but not Part 135 fixed-wing operations (NTSB, 2020a). This gap in safety has led to several fatal CFIT accidents involving flights operated under visual flight rules at low altitudes where TAWS alerts were inhibited due to the lack of adequate TAWS protections and nuisance-alert mitigations. Despite voluntary efforts to ramp up CFIT training among operators in Alaska, it was not a regulatory requirement at the time of this accident.

That meant that per the company's operational procedures, Hageland's CFIT training manual was a voluntary effort and not subject to FAA oversight. Hageland adopted training materials from the Medallion Foundation, which, in turn, developed its materials from the Flight Safety Foundation's 1995 CFIT training guidance—this training did not have applications for use with GPS.

For Flight 3153, both the PIC and the SIC had completed an online computer-based version of the CFIT training. However, there were no records that the SIC had completed any simulator training for CFIT avoidance.

The NTSB cited a hazardous brew of outdated materials, incomplete training for the SIC, and no FAA oversight in their investigation into Flight 3153. Furthermore, as highlighted earlier, the NTSB noted the discrepancy between Part 135 and Part 121 safety in their Most Wanted List for 2019–2020. Why does the FAA not require CFIT training for Part 135 operators like Hageland? Or, for that matter, a Safety Management System (SMS) or Flight Data Monitoring program? At the time of this accident, Hageland *exceeded* the minimum requirement for training on CFIT and related operational hazards.

Other operators in Alaska, who may be smaller or do not have the resources available to implement such a training program, are even more at risk for accidents related to continued VFR flight into IMC and CFIT. With the closure of the Medallion Foundation, which provided the CFIT training materials to Hageland, operators in Alaska may have to develop their own CFIT training or use third-party vendors.

Nevertheless, what are some of the causes of a CFIT accident? CFIT has been a longstanding problem in Alaskan aviation and, more broadly, in the Part 135 world of flying in the United States. CFIT is the leading cause of accidents in Alaska (Sumwalt, 2019), accounting for 80% of accidents in the state (NTSB, 2020b). The International Air Transport Organization (IATA) and other researchers suggest that the contributing factors are often attributed to human error, such as ineffective CRM, failure to follow established standard operating procedures (SOP), poor flight planning by the pilot, and inadequate flight path management (IATA, 2018).

Other factors are a lack of awareness of the aircraft's vertical and/or horizontal position in relation to terrain, unstable approaches, failure to initiate a go-around procedure when necessary, and adverse weather conditions. Lack of precision in approach and loss of situational awareness (S.A.) were also factors in CFIT accidents (IATA, 2018; Tomaszewska et al., 2020).

Failure to follow SOP may be a result of organizational drift. This theory is based on Scott A. Snook's theory of practical drift, which is used to understand how the performance of any system "drifts away" from its original design (ICAO, 2018). In this case, organizational drift occurs when the standard by which employees operate drifts from the procedures and policies as they were imagined at inception (ICAO, 2018).

The theory suggests that employees will look for ways to make their operations more straightforward and more efficient due to workload management issues with current procedures. The standards gradually erode with drift, and a new normal is implemented. However, continued success using these nonstandard procedures leads to more drift. This new normal and departure from standards may not be in the interest of safety.

There are scenarios where some tasks or job functions have inadequate or nonexistent procedures; this encourages individualized procedures to manage such tasks. In the case of Flight 3153, inhibiting the TAWS appears to have been part of the latter. The use of the switch was allowed. However, the training on when and where it should be used was inadequate, so pilots developed and used their procedures during flight operations.

The SIC had recently joined the company, and this flight was part of his second shift in the bush after completing initial operational experience (IOE) training with Hageland. Extant literature suggests that inexperienced pilots are more likely to give in to peer pressure from their more senior counterparts and may not voice safety concerns due to perceived or actual senior pilot authority (Michalski & Bearman, 2014; Noort et al., 2021). New pilots are also more susceptible to tolerating procedural drift from superiors. Behaviors with successful outcomes are more likely to be repeated, and the PIC had likely employed the TAWS inhibit switch in previous flights (Wiggins et al., 2012).

When asked about the safety culture at Hageland, the director of operations (DO) described their processes and culture as reactive but on the path to becoming a proactive organization. His perspective was that employees actively participated in safety and reporting programs—in short, they had a healthy safety culture (NTSB, 2018, p. 29). The lead pilot for the Bethel base was asked a poignant question in the post-accident debrief: *What is the biggest challenge of flying in Alaska?* His response? *Making the right decisions* (NTSB, 2018b, p. 26).

ORGANIZATIONAL CULTURE AND ACCIDENT CAUSATION

Promech (Ketchikan)

Ketchikan, Alaska, is one of the largest cities in Southeast Alaska. It is a popular cruise ship destination during the summer, and the population swells when cruise ships dock in the harbor. Tourism, which plays a significant role in the economy of Ketchikan and Southeast Alaska, is also a large contributor to the customer base for many local aviation operators who provide flight-seeing tours to cruise passengers. Competition is fierce in the Southeast, and many operators are vying for the same customers.

One summer day, June 25, 2015, a turbine-powered, float-equipped DHC-3 Otter impacted terrain during a flight-seeing tour near Ketchikan, Alaska, in the Misty Fjords National Monument. The pilot and all eight of the passengers were killed on impact. The occupants were passengers on a cruise ship transiting Southeast Alaska. The flight was sold as part of a "cruise/fly" package through the cruise line and air tour operator Promech Air.

For this tour, passengers would take a boat to a remote bay in the Misty Fjords National Monument. Once they reach this location, Rudyerd Bay, they fly back to Ketchikan Harbor on a float plane to complete the tour by both sea and air. The timing was essential, as the cruise ship would depart without its passengers should they miss the "all-aboard" time. If this were the case, the air tour operator would be responsible for arranging transportation to get the passengers to the next port of call.

Multiple tours were scheduled on this day, and things were running behind. Pilots could choose between two routes to and from Rudyerd Bay. The long route, primarily over water, was preferred when the weather conditions, such as visibility and cloud base, were lower. It took about 30 minutes to complete. Alternatively, the short route was more scenic and direct to Ketchikan, taking about 25 minutes and transiting the area over primarily terrain and inland lakes.

The weather was marginal VFR with cloud ceilings of about 1,000 to 3,000 feet and visibility of three to five miles. The weather forecasts showed minimal improvement throughout the day and even periods of IFR weather. On his first tour of the day, the accident pilot departed in a group of three aircraft. He was the first pilot to depart and reported good weather through the short route to his colleagues. The other two pilots elected to take the long route on the outbound leg. Returning inbound to Ketchikan, all three pilots attempted the short route, but all ultimately diverted off it to take the long route back instead. The accident pilot was again first to depart—his flight path showed several course reversals and altitude changes to deviate around the weather.

The accident pilot was the fourth of six airplanes to depart Ketchikan on the day's second tour. Each aircraft flew the long route and had to work with weather issues along the route. All but one pilot descended below 500 feet above ground level to avoid clouds. One pilot in the group of six expressed concern about the weather before departure yet ultimately flew the tour. This pilot expressed relief to be done with flying for the day when he returned to Ketchikan. The company president/CEO also flew the second tour and, along with the accident pilot, ended up flying below 400 *ft* AGL.

On the third tour, four aircraft made it out to Rudyerd Bay without incident. The passengers had a 12:30 all-aboard time for their cruise ship and were scheduled to depart Rudyerd Bay at 11:45. Running behind schedule, Aircraft

One and Four elected to take the short route back to Ketchikan, and Aircraft Two took the long route. The accident aircraft departed number three, did not leave until 12:07 pm, and attempted to take the short route back to Ketchikan.

Promech Air began a search for the aircraft after it failed to return to Ketchikan Harbor. With aid from an emergency locator transmitter signal, a helicopter from another area operator was able to locate the aircraft once the weather lifted. It was confirmed to have crashed into terrain near Ella Lake, a lake on the short route back to Ketchikan Harbor. Later that afternoon, search and rescue volunteers were able to verify that all passengers and the pilot had been killed in the accident.

While on the short route back to Ketchikan Harbor, the pilot entered Ella Narrows, a 100-to-300-foot valley with surrounding mountains between 2,000 and 2,500 feet. Data recovered from the aircraft showed that he entered this area at approximately 1,300 feet, rapidly descended to 1,000 feet, and climbed back to 1,300 feet. Transiting over Ella Lake, pilots would typically pass two similar-looking mountains before starting a slight right turn to continue their flight.

The pilot passed over a ridge on the lake's western shore at 1,500 feet, maintained heading, and two seconds before impact, the aircraft recorded 2+Gs on the aircraft due to the pilot's rapid upward pitch. The pilot had started his turn too early, only passing the first of the two mountains. He would have cleared all terrain at his selected altitude if he had continued down the lake slightly longer.

Photographs from passenger cell phones show low clouds, fog, and obscured terrain prior to impact. Weather data from the National Weather Service predicted isolated low ceilings below 1,000 feet for the area, visibility between three and five miles, and light rain with mist. Mountain obscuration was forecast in the AIRMET, and TAF data for Ketchikan International Airport forecast lower ceilings of 1,000 feet. An FAA weather camera facing the direction of the accident site showed ceilings obscuring the top of the terrain in the area.

Using photographs from passengers and interview data from the other pilots that day, it was determined that the accident pilot likely encountered visibility between one and two miles and was flying through an area where terrain was obscured above 1,000 feet. In the Ketchikan area, numerous other operators canceled their flights for the day, and few still launched aircraft on air tours.

Analysis

An accident like this does not just happen because one pilot chose to fly in marginal conditions. The NTSB investigation into this accident revealed many latent factors within the organization that did not trap errors that led to the death of this pilot and his passengers.

First, adequate training. The CFIT avoidance training program was voluntary at Promech at the time of the accident. Ground and simulator training was available to pilots. However, other operators questioned how beneficial the simulator training was, arguing that nothing was comparable to actual experience, with real distractions and real task saturation. The DO and another company pilot had differing opinions of the accident pilot's skill during CFIT escape maneuvers. Additionally, the pilot had recently transitioned from the DHC-2 Beaver to the DHC-3 Otter, a more powerful and capable aircraft. Other pilots at the company expressed concern about his readiness for this transition.

If there were questions about the pilot's ability to operate the aircraft safely, why was he not given extra training? The answers to this may lie within the safety programs at Promech at the time.

Without a formal safety program, Promech relied on verbal reports directly to the DO or written reports submitted to a physical drop box in the employee lounge. In such a small company, it seems unlikely that anonymity, a key component of modern safety programs, could be assured. In light of this, management was unaware of significant recent safety events at the company.

The accident pilot struck trees with his floats while landing on a previous air tour. On the day of the accident, management was unaware that the pilot had turned around due to poor weather and that two other Promech pilots were forced to find alternate routes around the weather the morning of the accident. Open communication—without fear of retaliation—goes a long way to fostering a proactive safety culture in an organization.

Additionally, there was some confusion about the safety program in the post-accident interviews. Pilots could not give confident answers about who was responsible for the safety programs at Promech. There were also differing opinions on the safety culture. Some pilots felt comfortable discussing safety concerns. Others, however, described situations in which they felt pressured by management to fly in marginal weather conditions—they were taught (through company culture) that bending the rules was synonymous with flying in Alaska. This attitude had become normative for the organization (and many others across Alaska). It was assumed that the flying culture in Alaska was "we push through, we push through" (NTSB, 2017a, p. 27).

Pressures to perform like more experienced pilots likely played a role in the pilot's decision to fly that day, and he had no backup from the flight scheduler, who had operational control of his flight. She was a longtime employee of Promech yet had not been given any recurrent training since her initial onboarding. Because only limited operational control could be given to a flight scheduler, it was a requirement that the pilot and scheduler both converse and agree that conditions were acceptable to dispatch a flight—either could terminate a flight at any time due to changing weather conditions. On this day, there were no discussions between the two employees outside of pleasantries and one weather report from the outbound leg of the first flight.

Launching in groups of multiple aircraft gives the illusion of safety in numbers, especially when the company president/CEO is one of the pilots on a tour. For management to willingly fly lower than VFR minimums, it opens the door for other company pilots to assume this behavior is acceptable.

It is worth mentioning that the aircraft involved in the accident was equipped with a TAWS. Much like the Togiak accident discussed in the previous section, this switch was selected to "inhibit" during the post-accident aircraft inspection.

THE PRODUCTION AND PROTECTION CONUNDRUM (EXTERNAL PRESSURES)

Ketchikan Mid-air

The Misty Fjords National Monument is a popular destination for tourists embarking on flight-seeing tours while traveling through Alaska on their cruise vacations. On May 19, 2019, two such aircraft collided near a waterfall in the Monument, approximately 8NM to the N.E. of Ketchikan. The aircraft, a DHC-2 (Beaver) and a DHC-3 (Otter), were both operating under visual flight rules in VMC conditions at the time of the accident (NTSB, 2021). The Beaver was wholly destroyed in the accident, and the pilot and all four passengers were killed. The Otter sustained significant damage—one passenger was killed, nine sustained serious injuries, and the pilot sustained only minor injuries.

The flight path data recovered during the accident investigation showed that both aircraft departed Ketchikan at approximately the same time, shortly after 1213 p.m. The aircraft converged on the waterfall from different altitudes: the Beaver began a climb from approximately 3,176 feet, and the Otter descended from 4,000 feet. The two aircraft collided at 1221 local time at 3,350 feet.

By analyzing the wreckage from both aircraft, the NTSB determined that the Otter had impacted from above and behind the Beaver due to the mechanical cuts that tracked inboard from midway on the right wing. This wing was separated from the aircraft at the wreckage site, located both on land and in the water near the collision. All of the passengers were ejected from the aircraft.

The Otter also sustained substantial damage from the collision and its subsequent impact with water. The propeller, which caused significant damage to the Beaver, had damage consistent with rotational gouging. On crash landing in the water, both floats separated from the aircraft, and the fuselage broke into two parts, both of which managed to stay mostly intact.

The pilot of the Otter was able to swim out of the wreckage and help assist the nine surviving passengers. The tenth passenger was sitting in the co-pilot seat, and her cause of death was blunt force trauma, with drowning as a contributing factor. The passengers were briefed on the location and use of the life vests, yet only one took one when evacuating the aircraft after the crash. The survivors were assisted by a passing good Samaritan and taken to the hospital.

Analysis

Both aircraft were operating under VFR in VMC conditions. Clear skies were reported at Ketchikan airport. On such a beautiful day, the pilots relied primarily on "see and avoid" techniques as they transited through the airspace in Misty Fjords National Monument.

In the aftermath of this accident, the NTSB used a "cockpit visibility study" to analyze the field of vision of both pilots using passenger photographs, aircraft specifications, and sun angles. The objective was to determine if either pilot had the opportunity to see the other pilot before they collided mid-air. Their analysis: no. There was not enough time for either pilot to see the other aircraft by visually scanning outside.

Based on the flight paths, the pilot of the Otter would have had approximately 1 minute and 20 seconds of opportunity to see the Beaver before impact. Of course, this assumes that he would have been able to see something the size of a thumbnail held at arm's length. In addition, most of the Beaver was concealed by the window post until impact, making a positive identification of the Beaver near impossible. Relying on the see and avoid concept was unrealistic in avoiding this accident.

ADS-B, which was installed in both aircraft, was on and transmitting. These systems use airborne information from the aircraft to transmit position data to ground stations and other aircraft equipped with ADS-B and help avoid traffic conflicts. The information from the Otter was missing altitude data due to a switch selection. The pilot was unaware this was the case—his preflight routine was to leave it in the last position, assuming that moving the selector would interfere with the avionics equipment. Despite this, he said the display was functional, and he saw other traffic on the display during the accident flight.

The NTSB has underscored the effectiveness of using devices that display traffic information in the cockpit. They have also emphasized the limitations of the see-and-avoid concept in this and previous accidents. A cockpit display of traffic information (CDTI) effectively draws pilots' attention to potential traffic conflicts in high-traffic tour areas, such as the Misty Fjords National Monument or the Grand Canyon. This is especially true when the CDTI is coupled with aural and visual alerting systems while the pilot has a high visual workload. If used as part of a normal scan, the pilot can maintain situational awareness of other traffic in the area, especially when operating a single pilot.

The takeaway from this analysis is that safety controls should always be layered, and the rule of redundancy is essential. The human's visual search outside the cockpit view can be hindered by high workload and distractors. Attentional issues can affect scan patterns, leading to the inability to detect conflicting traffic even under high visibility. That is why the complementary role of technologies such as ADS-B and CDTI is desired in Part 135operations. Effective training that also highlights the limitations of such technologies should be highlighted to reduce the unintended consequences of attentional tunneling and over-dependency.

5 Understanding Pilot Decision-Making in Alaskan Operational Settings

A Qualitative Approach

So, how do you study a group of people who live in a remote place in the far corner of the United States? The answer: live there yourself! It would be unfair of us not to acknowledge the difficulty of conducting an empirical study in such a place. Given the prevalence of technologies like video chat and other web-based applications, finding participants and building a rapport with them is challenging for any qualitative researcher.

For this book on understanding the decision-making processes of Alaskan pilots, we used a qualitative research approach with a combination of phenomenological and ethnographic methods. We felt this was the most appropriate means of conducting our study due to its emphasis on the cultural themes of the participant group—especially given that there has been little empirical research on pilots in the state of Alaska in recent times. Our study aimed to understand the factors influencing pilot decision-making in remote Alaska.

DOI: 10.1201/9781032678863-5

The focus was also to explore and understand the lived experiences of these pilots in the Alaskan aviation environment and how it influenced their operational decision-making.

Having lived and flown in Alaska, Dana navigated the sometimes-confusing landscape of the different types of operations that take place there. With his experience flying in some remote areas of Ghana in West Africa, Kwasi provides another perspective that adds context to the readers' understanding of remote-setting environments, challenges, and decision-making styles. We used a semi-structured interview methodology to elicit the opinions of our research respondents who volunteered to be part of the study.

INTERVIEWING

Interviews are a vital component of many qualitative research studies. In exploratory studies, where it is important to understand the lifestyles and feelings of participants, interview data is crucial to unearthing perspectives and gauging opinions, which assist in understanding various phenomena associated with an inquiry. In this endeavor, we sought to interview our participants in their remote working environments.

Interview data has been paramount for many studies on aviation in remote settings, especially those in Alaska. Finding participants may pose a challenge—how can a researcher talk to pilots who may self-disclose potentially safety-sensitive information into the public domain after publishing our findings? How can we glean essential information from these critical stakeholders to help inform policies and practices to improve aviation safety? As is evident to anyone who works in a niche field, professional networks are relatively small, and it helps to leverage these to find participants.

Another challenge when interviewing pilots is where to start. Using a semi-structured interview was beneficial to maintain consistency between participants. We asked about situations they may encounter in daily flight operations and attempted to understand their beliefs and experiences about flying in Alaska.

By keeping the core questions the same between participants, we were able to gather meaningful data about their beliefs and experiences while also having data that covered the same topics between interview participants. Creating a draft of interview questions and beta-testing it on someone familiar with Alaskan aviation operations was a great way to determine its usefulness. It also allowed us to draft follow-up questions that helped to guide the conversation and uncover any unexplored areas of remote aviation operations.

THE ETHICAL DILEMMA

Notoriously, pilots love to talk. Whether it is about their latest work trip, the newest aviation gadget on the market, or just plain hangar talk, sometimes getting a pilot to *stop* talking seems complicated. In a casual setting, pilots will retell stories about flying throughout their careers, including narratives where behaviors and outcomes may not have been strictly legal.

Alaska is a small community, and the pilot community is even smaller. The interviewer must remember that maintaining the privacy of a pilot being interviewed is very important to them. There are few degrees of separation, so removing identifying information from interview data is critically important. The community is so close-knit that it could be possible to identify a person based on relaying information about the type of aircraft they flew and the general location they frequented.

The contents of this book were based on graduate thesis research (Atkins, 2021). We disclosed to the participants as part of a statement of consent that the study's findings could be used in further publications to add to the existing literature on aviation in Alaska while providing data-driven recommendations for safety improvements in Alaskan aviation operations. Qualitative research in this domain intends to gather information to understand the central phenomenon of operational complexities and the decision-making styles of pilots who fly in Alaska.

The information gathered in this study was solely to inform policies and practices to improve aviation safety. Because of that, confidentiality was essential, and every attempt was made to ensure the participants' privacy and the source data's confidentiality. Emphasizing confidentiality was also a means to assuage any fear of potential reprisals for comments and engender trust between the researchers and the interview participants.

QUALITATIVE DATA ANALYSIS

Our research started with a documentary review of the statistics (and from firsthand exposure) of aviation accidents in Alaska. This type of qualitative data analysis was the foundation for the research interview protocol—we could look at important causal and contributory factors the NTSB had identified in their accident reviews and derive questions to ask our participants regarding these themes.

Content (or document) analysis examines aspects of published documents and data such as "frequency, type, correlation, and absence" (Saldaña & Omasta, 2018). This type of data analysis is commonplace in quantitative research. In the qualitative world, deriving themes and common categories from the documents makes up the content analysis. We looked for themes like personnel, organization, or environmental factors contributing to accidents (this is also how the NTSB codes their accident factors). After reviewing the data, we asked ourselves the following questions:

- What is the objective of the author in publishing the accident reports?
    - In this case, the accident data came from the NTSB database. Their objective is to provide a non-biased analysis of accident contributory and causal factors to improve aviation safety.
- What are the recurring themes we are seeing in the accident data?
- What is important to ask participants based on these themes?
- How would participant perspectives help change our analysis?
- How would participant perspectives help answer questions raised due to our document analysis?

For us to do a comprehensive ethnographic case study and follow our content analysis of accident data in the state, we felt it was important to choose a few accidents and review them in detail. We chose the three case studies you read about in the previous chapter based on their recency, impact on the aviation community in the state, and their availability of comprehensive review from the NTSB and FAA. These case studies helped us develop our interview protocol as the primary research instrument.

CODES AND THEMES

Using the interview data, we could categorize participant perspectives into different themes. Data were first coded using value and descriptive methods and then compared. As pilots ourselves, coding the data came more naturally than someone who may have had less extensive flight experience. This common experience shared with interview participants was beneficial in two ways: it allowed us to build rapport with interview participants quickly, and it allowed us to code the interview data using a pilot lens. It is worth noting, however, that our experiences as pilots were the most obvious place for bias to be introduced into the study.

We chose to use value and descriptive methods to code our data. We chose these methods due to their cultural lens on coding, which emphasizes participant experiences and actions in intrapersonal and interpersonal ways (Saldana & Omasta, 2018). The descriptive methods highlight the day-to-day actions of participants and classify their physical environments. The codes we found were a combination of more concrete descriptions, like their work environment, and more abstract ideas, like their attitudes and feelings toward safety. We then could group the significant codes that stood out across all our interviews into six major themes. These themes made up the bulk of our analysis.

The coding style highlighted earlier is particularly appropriate for investigating the aviation world, especially given the human factors of flying in a remote setting. The environment itself creates unique challenges, all of which were attested to by interview participants (descriptive coding). The *human factors vs. decision-making* element of our participants' perspectives lends itself well to a value coding method. It is important to understand both. To use a common trope, the man, myth, and machine are equally important to understand in this context.

6 Human and Organizational Factors Influencing Aeronautical Decision-Making in Remote Operational Settings

Real-Time Perspectives

When we looked at how best to research pilots in Alaska, we determined that using a qualitative study was the best fit. Hangar talk, sharing stories, and sharing the delights of flying in Alaska are best done in person, in a casual setting. We wanted to emulate that for our study. Unfortunately, global pandemics

DOI: 10.1201/9781032678863-6

(COVID-19) throw off even best-laid plans, so our interviews were done using the audio/video platform Zoom. However, it was still important to us to recreate a casual environment to the best of our abilities, and we achieved this by using a semi-structured interview process.

Our interview protocol and questionnaire were developed after a thorough case study analysis and a deep dive into the NTSB data on accidents in Alaska. We looked for patterns and themes recurring in the data, NTSB recommendations to operators, and read through causal and contributory factors to understand what was causing accidents in the state.

These analyses also shed light on the areas where further qualitative analysis was required to help describe and explain the psychosocial and organizational factors influencing Alaskan pilot decision-making. Afterward, it was up to us to craft an interview guide that touched on those points while allowing the interviewee to share their perspectives, which was the study's primary objective.

We found a series of recurring themes that were discussed repeatedly across our group of participants. Eight pilots volunteered to participate in our study, and we were excited that our participants had a variety of backgrounds and experiences. The most experienced had been flying in Alaska for over 20 years, while the newest was working their first commercial job. This range of experience was essential to us—pilots' perspectives at the beginning of their careers were just as important as those of pilots close to retirement age.

The questions we developed for our semi-structured interviews were beta-tested on a colleague who flew part 135 in Alaska while building his flight time before transitioning to the airline world. Draft follow-up questions were created as a result of that test interview. It was a good practice run and allowed us to determine how best to create a causal environment when talking with our actual participants.

Six themes became apparent, which we will discuss in depth throughout this section. Transcribing, coding, and doing the thematic analysis of the interviews was a lengthy process, and surprisingly, we elected to do it by hand. This grueling process was made easier by the fact that we could get very familiar with the interview data, know what segments needed to be edited to remove any identifying information about the participants, and start to get a feel for what the overall attitudes toward flying in the state were among this group.

Our first major theme was pilot relationships with three salient codes and the definitions depicted in Table 6.1. The highlighted coverage is a descriptive statistical output of the percentage of counts or textual output representative of the code as compared to the overall textual output for the theme. We derived the textual counts using an Excel spreadsheet during the manual coding process.

PILOT RELATIONSHIPS

TABLE 6.1 Theme 1: Pilot Relationships

CODE	*DESCRIPTION*	*COVERAGE*
Camaraderie	Developing friendships, trust between colleagues, mutual respect for others' boundaries/decision-making, team dynamics	75%
Cockpit relationship (CRM)	Relationship between a captain and a first officer, cockpit relationship dynamics, superior/inferior power balance	87.5%
Peer pressure	Positive or negative influence on decision-making	50%

Insights from the airline world all point to the significance of developing good and effective CRM between pilots in the flight deck. Understanding these relationships was paramount as part of this study. Relationships became resources for pilots to use during their decision-making. Conversely, we found that pilots could also be significantly influenced by other pilots who were in their environment—from both a positive and negative lens.

Our participants discussed the importance of CRM even in companies that operated solely with single pilot operations. The "crew" was expanded to include pilots in other aircraft, dispatchers and flight followers on the ground, and non-company personnel who gave information to the crews (such as those who provided weather and runway condition reports). The dynamic of a crewed flight deck completely changed the flight experience and how decisions were made. Next, two participants discuss the importance of relationships with their pilot peers:

Quote 1

> *I made a lot of really good friends, and a lot of these guys . . . We rely on each other. Whoever goes in there first, we're relying on a PIREP [pilot report] . . . We also make decisions together. There are a few pilots I legitimately trust. We'd be talking on the radio on the way down to who knows where, all these little, tiny villages, which may or may not have an approach. We're discussing the different options based upon the weather and everything we saw this morning or the morning of the flight. Try to figure out what the best way was to get in there. Talking to each other is huge. But then there are other pilots that I never ask an opinion of.*

Quote 2

> *I really love the attitude where you're a team. It's not, I mean it is a captain and an FO [first officer], but I feel for me like it's more of a team. I had some captains like that, and that was amazing. And then I had some captains that were like "I am the captain, you're just a warm body in the seat" and I know how that made me feel.*

Working together outside of the flight deck is essential in the "bush country" of Alaska. Pilots are frequently on shift together in remote locations across the state. They work together, live together, and spend their free time together while they are away from home. This dynamic plays into the social theories we discussed earlier. To minimize conflicts, it is essential to have compatible personalities. This example highlights how social identity theory plays into bush environments. You are either in . . . or you are out:

> *In bush Alaska in 135 operations, the nature of working with and around the same people all the time . . . . You either fit into the team or you don't last. People, this is a harsh thing to say, but the people who can't get along in that circumstance and who can't contribute get run off.*

Perspectives like this highlight the role of peer pressure and group effects on group decision-making dynamics. Fitting in while on shift in bush Alaska is essential to have a positive experience while at work away from home.

The relationships pilots form with their peers influence their decision-making. Positive relationships make good pilots; these could be relationships that encourage safe decision-making, canceling flights where appropriate, and incrementally increasing one's exposure to risk. Other pilots discussed how the behaviors of their peers have negatively pressured them to fly outside of their comfort zone. New, inexperienced pilots are more susceptible to pressure from operationally experienced pilots. This is a quote from a respondent:

> *I think one of the things I've seen that's problematic in places that I go that have larger pilot bases was a classic example where we had 14 pilots on shift. We had some really experienced pilots, and then we had people that were on their first commercial job. It seemed like one person was deciding for that whole base that it's flyable. "We're good everybody. Everybody to your planes." Whereas after, I went where there was one [Aircraft Type A] driver, and one [Aircraft Type B] driver. I was paired up with an experienced driver and I was a fairly new commercial driver, but that was better because I didn't feel the pressure of these other drivers all running to their planes.*

We found that the consensus among our participants was that pilots who knew their boundaries and stuck to them when flying off-airport or in challenging weather conditions were considered the best bush pilots. Our most experienced pilot indicated that safety in the Part 135 flying world often depends on a combination of a few factors: a pilot's manual flying skill, their ability to interpret their environment (e.g., a pilot on floats needs to be able to read the water), and their willingness to gradually expand their flying skillset by incrementally increasing the challenge level of their operations.

Negative peer pressure is commonplace, too. Comparing skills to other pilots in the workplace is not always bad, but assuming that conditions are safe or that an appropriate level of risk depends on the individual pilot—just because one individual thinks it is safe does not mean it actually is. Sticking to one's convictions about one's piloting abilities is crucial to ensure that risk is managed appropriately. A respondent provided this interesting perspective:

> *You have to understand risk management, you have to understand your limitations and you have to be willing to stand up for your limitations. Don't be pushed into something. It's one thing to expand your knowledge base and to work on your craft but it's another thing to go beyond what you feel is safe just because someone else feels it's safe.*

A good understanding of a pilot's limitations when it comes to skills and expertise hinges on attitudes, and that is what we explore in the next theme, which is highlighted in Table 6.2.

PILOT ATTITUDES

TABLE 6.2 Theme 2: Pilot Attitudes

CODE	*DESCRIPTION*	*COVERAGE*
Ego	Sense of self-importance, boastfulness, comparison against other pilots	87.5%
Confidence	Comfort in skills, self-pride, self-assurance	87.5%
Risk tolerance	Pilot's willingness to accept risk, to accomplish dangerous tasks	87.5%

In an earlier chapter, we discussed the influence of personality and attitude from an academic perspective. Personality, being relatively constant throughout a lifetime, could draw a person to a career in aviation (Fitzgibbons et al., 2000). On the other hand, attitude is a much better determinant of how successful a pilot will be throughout training and operations on the line. Ego and confidence can be commonly mistaken for one another, and it was these two similar but different attitudes that participants in our study discussed as make-or-break.

How does a pilot build confidence? That circles back to their attitudes toward their jobs, training, and skills development, which we will discuss later. As much as confidence honed by technical knowledge and proficient skills are good in aviation, *overconfidence* can be detrimental to effective decision-making.

Overconfidence bias is when an individual overestimates the state of their knowledge and ability to perform well (Chira & Adams, 2011). In any given situation, this bias can influence pilots to rapidly decide on a course of action without considering alternative hypotheses. They may not consider other crew members' opinions, limiting decision effectiveness. Overconfidence can lead to cognitive tunneling, which is when a pilot's ability to detect relevant cues from their environment is hampered—they have chosen a course of action and are no longer considering additional information (Carroll, 2023; Wickens et al., 2021).

Overconfidence can be a primer for errors on the flight deck, and the outcomes are more extreme in this remote environment. It was interesting to probe into the mindset of our interview pilots and ask about overconfidence and other at-risk behavioral patterns. One of the pilots stated:

> *The pilot is required to have an attitude of "I want to broaden my horizons; I want to push my boundaries." But then a cockiness or an overconfidence or a feeling that I can do anything is taking that too far and that can have disastrous results as well.*

It seems that ego is acquired naturally by pilots who fly in the bush. It is an element of working around the same people, building relationships, and fitting into the in-group while working in extreme and remote settings. Our participants discussed how ego was part of the job—it evolves with experience.

> *There is a part of it that's just pilot ego. Each one of us has a touch of that if we're honest with ourselves. The longer I fly the less I feel like I have. The more you know, the more you realize you don't know.*

Ego and negative attitudes toward flying can erode relationships between pilots. Inversely, positive attitudes can help foster camaraderie and sound decision-making. Attitudes show themselves in all sorts of settings. Hangar talk provides a great platform to understand and assess remote pilots' general disposition toward various philosophies of flight safety.

Hangar talk is the industry jargon for informal hang-outs where pilots share stories about their flight experiences in a non-judgmental environment. It is akin to a "talking shop" with minimal formal moderation. Unsurprisingly, it is an environment where uninhibited attitudes and egos may manifest themselves most readily.

It is also an environment where pilots share lessons learned in a positive light. It is unsurprising to hear stories of pilots talking about episodes of near-fuel exhaustion or harrowing encounters with inclement weather conditions such as wind shear. With the advent of technology, some hangar talks have transitioned into podcasts, but in the remote settings of Alaska, the old-fashioned physical hangar talk is quite common.

A pilot displaying egotistical tendencies might share these stories in a braggadocious manner. They may elicit an ego boost from the reactions they get from their peers. Others take a different approach of being contemplative and using the story as a learning event for improvement. One of our participants explained the difference between sharing stories with the intent of boosting themselves and sharing experiences to inform the pilot group better:

> *Stories are kind of like an ego boost. Bragging, "This is the weather I flew in VFR today, you'll never believe this," "This is how much ice I built up on this flight I couldn't believe it." Then the sharing experiences is more like, "Hey, you know I had this failure today, this type of system failure," or "I had this happen on touchdown," or "this mountain at a certain airport causes this type of wind pattern to come over so watch it when you go in there, I haven't seen it before."*

One major highlight of our study was that our participants noted that the overwhelming majority of pilots who fly in Alaska have a positive disposition and attitude toward their work. They genuinely want to use their learned experiences to teach or mentor the next generation of pilots. This helps build the camaraderie crucial to creating a positive work environment, creating a safety-conscious pilot group, and creating pilots with well-tempered aeronautical decision-making. Here is a candid example:

> *I'd say it's positive. It's not a dick-measuring contest . . . . There are people like that, but they get run off. People just kind of say, "Hey watch out for this guy" or "Whatever, he's a blowhard." The vast majority of those stories it's . . . we don't want to have to see someone else learn the hard way.*

Ego, confidence, and risk tolerance are all elements factored into the decision-making matrix each pilot must consider before each flight. Being a safe and successful pilot comes from a commensuration of risk tolerance with the operational benefits of flight missions. Playing it "too safe" is unrealistic; blue-bird days, those without a cloud in the sky, are a commodity, not a commonality. One participant sums it up succinctly:

> *Unfortunately, if you wait for huge margins or everything, you're not going to make it work half the time.*

On the other end of the spectrum, flying in this remote setting of Alaska without regard for the risk posed by safety hazards can result in undesired operational states. As we have seen time and time again, fatal outcomes are commonplace in Alaska. If pilots accept flights in poor weather conditions or they accept flights that push far beyond the limits of their piloting abilities, we might be reading about them in the newspaper the following day. This was a statement from one of the respondents on the excessive risk tolerability of a pilot involved in an accident:

> *He was the epitome of a cowboy. He was on the extreme edge of that. I don't think most people fall to the extremes one way or the other. He was the exact . . . like when you heard that he had run into the side of the mountain in bad weather it was like, "it was only a matter of time."*

Fortunately, most pilots within our population in Alaska are neither prohibitively risk-averse nor dangerously accepting of flights. Instead, they fall somewhere in the middle. It is critical to success to find the balance between playing it unnecessarily safe and taking all flights regardless of internal and external risk factors. Most of our participants reiterated that pilots must display professionalism and good judgment by assessing their flying skills (manual and cognitive) with the limitations posed by variables like weather, aviation infrastructure, and aircraft capabilities.

Consequently, this means that a particular pilot would sometimes be more appropriate for a flight due to their experience, skill level, or familiarity. A pilot respondent had this to say about skills levels and comfortability with risk.

> *It's crazy to me to hear people who go in there day in and day out versus people who still go in there a fair amount—how their approach to it is different. When I'm hearing these things and weighing "How comfortable am I with this airport?" it makes you think. I've turned down a few flights in the Southeast that maybe someone else wouldn't have. Not because he could do it any safer than I could, but because he just knew a little more what to expect. So that's an interesting point, and I think that's true in a lot of different places in Alaska. If people are comfortable going into that place, they're accepting that flight.*

SKILL ENHANCEMENT

TABLE 6.3 Theme 3: Skill Enhancement

CODE	*DESCRIPTION*	*COVERAGE*
AK experience	Experience specific to flying in Alaska. Much like "instrument time" refers to time spent with reference to the flight instruments, "AK time" references time flown in the state of Alaska.	100%
Mentorship	The relationship between a pilot and a mentor, especially someone respected in the company or the aviation industry	62.5%
Pilot development training	Company-provided training; professional development provided by the organization. This could include ground school, computer-based training, initial operating experience (IOE), and training on the flight line.	62.5%
Skill enhancement	An individual's outlook toward training and improving, personal growth/career development, attitudes required to learn.	50%

So, how does a pilot become a *bush* pilot in the first place? Results from our interviews describe how the methods by which pilots gain experience are crucial to their safety as pilots in the state. These methods are summarized in Table 6.3. The vast diversity of terrain, extreme weather, and the intensity of flying in Alaska all play a part in this. Good experience is the "how" to become successful over a long career in the state of Alaska. The best approach is to gain that experience incrementally and learn from other experienced pilots.

> *Experience is the only thing that is truly going to make a bush pilot. You come up here and no matter how good a stick you are, there's nothing that can prepare you for everything that Alaska is going to throw at you until you experience it. That's why I think having mentors, having people who you can fly along with before you get thrown into the deep end by yourself, is so crucial. There are many, many instances where I just learned from trial and error, or I experienced things by myself for the first time.*

While interviewing our participants, it was extremely relatable to hear that they felt unqualified when they first started flying in Alaska. When I (Dana) first moved to Alaska, I thought I knew how to fly. I thought I knew how to evaluate weather conditions and plan flights. I thought I was proficient in my manual flying skills (stick and rudder). Truthfully, I was incredibly naive to the challenges of flying in Alaska and was quickly humbled by the reality of operating in such a remote environment!

It felt like there was a chasm of skills between me and my coworkers, who had been flying here much longer than I had been flying, period! One participant took the words right out of my mouth:

> *Looking back on my starting years, I really had no business flying in Alaska . . . coming in from the Midwest and doing long flat flights to nice airports with nice lighting. But I had a lot of multi-time, and that's the only reason I got hired up in Alaska . . .. I really had no expectations for what I was getting into, and the experience is a must. I've seen that kind of bite people at some established companies.*

I also had no business flying in Alaska! Luckily, I was able to work with some incredible pilots and learn along the way. Maintaining a positive safety record comes down to a few things, such as honesty about prior flying experience. Our participants were more willing to help new pilots who came in and were transparent about their experience.

New pilots must be willing to learn new skills (both manual flying and different decision-making techniques). It is cyclical; the more willing someone is to learn, the more they *will* learn; the more they learn, the more skilled they become, and then the more willing they are to learn even more.

Top-level management of air service operations should take a similar approach and consider the flight experience of a potential pilot hire to determine how best to integrate them into the culture at a given company. A pilot with no experience flying in Alaska should not be expected to perform the same tasks as a pilot with previous Alaska experience until they develop skills relevant to the environment and the operation. These pilots need to be weaned into flight operations incrementally, gradually gaining exposure to new and more challenging operations.

This strategy of incremental exposure helps build confidence, which makes pilots safer and more self-aware. Some of the actions management can take to implement this philosophy include increasing time spent on initial operating experience (IOE), pairing new first officers (FOs) with safe and experienced captains, gradually giving more challenging flights to the pilot, and reviewing safety hazards associated with the operation with the pilot.

As part of human capital development within an organization, training, and skill development are crucial to building both expertise and experience. Effective and expeditious performance feedback on training flights is critical to enhance safe experience building and it is vital in minimizing safety events such as accidents and incidents. A respondent in the study noted:

> *If you take a brand-new pilot, and you send them on one of the most challenging operations, it's almost a guarantee that they're going to wreck the airplane. Maybe not on that one, but if that's your operational philosophy, that any pilot can do anything that we have regardless of if they're brand new or they've been there forever . . . I'll just restate what I've said before, I've seen that result in a lot of accidents up here.*

Attitude is a great predisposition for learning operational skills. Alaska is challenging and unique, and a thirst for new knowledge is how a pilot will learn. Our participants thought that satisfaction with mediocrity and an unwillingness to improve skills and knowledge further may be operationally detrimental to pilots and detract from their ability to form camaraderie with other pilots. A new pilot willing to learn something new and enhance skills will be on the right path to improve and succeed in this setting.

> *You have to be willing to listen to other people and gain experience slowly and methodically, but also push your boundaries.*

Another equally important factor was a pilot's readiness to learn from the experiences of other pilots. The concept of mentorship repeatedly emerged while we talked to participants. Mentorship is essential in developing individuals both personally and professionally. It is a type of professional relationship in which a person with greater experience and wisdom guides another person toward enrichment (Air Force, 2019). This study explored how a mentor's experience, expertise, and professionalism are shared with the mentee in a mutually beneficial relationship.

Mentorship (formal, casual, or otherwise) significantly impacted our pilots. When asked about the most respected pilot type, four participants described pilots who were keen to share experiences through teaching and coaching. A core attribute was that their mentors would share all experiences, even those that were embarrassing or unsafe, to teach their mentees. No formal mentorship program was available to any of the participants; instead, a natural bond formed between pilots keen to pass on knowledge.

Using the experiences of others is an extremely valuable method of gaining knowledge. Openness, honesty, and willingness to share experiences where pilots made mistakes were especially valuable traits for young pilots to see in their coworkers. Positive hangar talk sessions were a powerful tool for creating camaraderie among pilots, learning from second-hand experience, and building confidence.

As pilots develop their skills and start to "learn the ropes" of flying in remote settings, they need the advice of others less and less. They get "kicked out of the nest," so to speak. At this stage, if the learning outcomes of the mentorship process have stuck with the pilot, they will begin to share their own experiences with the next generation of pilots. This mentorship cycle can help transform bush pilots' social and professional culture. If the attitudes steadily shift toward a safety-minded outlook and camaraderie and openness are encouraged, safety culture can gradually become a core part of flying in the bush. A respondent with considerable flying experience in Alaska provided context noting:

> *I think for us up here there's a lot of respect for the people that have experience operating up here. You hear someone's story, and you hear their experience and you're like, "Oh yeah I wouldn't have thought of that before . . ." They either reiterate what you already knew or cause you to think in a way that you probably hadn't thought before . . . . In general, there's kind of a wealth of knowledge that's open and available. I'd say from [the beginning] to where I am now, it's slowly decreased just because the need has decreased. Now, I feel like I find myself on the other end of the spectrum, sharing knowledge that you've learned either from other mentors or from firsthand experiences or mistakes.*

If a casual relationship between two pilots can have such a positive impact on skill development and learning in a remote setting, formal education in the form of company training should be even more influential. On-the-job training (OJT) seems to be a valuable complement to formal classroom training.

This OJT concept was corroborated by our participants, who seem to like the incremental exposure to more challenging flight environments as part of the OJT. As part of the new hire indoctrination training in Alaska, the FAA requires a Part 135 company-sponsored ground school to cover topics such as aircraft systems, company procedures, and operations specifications. Afterward, new hires complete flight training, where they learn how the company wants them to operate their aircraft.

Among some air operators, formal ground school is integrated with the initial operation experience (IOE) where aircraft technical systems and operational procedures are explained to new hires by the training captains during revenue flights. The following is a quote from a pilot new to the Alaskan setting:

> *My training wasn't very involved. The check ride wasn't very involved. I don't know if this is common in Part* 135 *operations. It's kind of like the check ride was the first training flight. "Here's how you do this and you're ready for the test and now you start learning." They kind of throw you into it.. . . Other than that, most of the training and the flying all happens on the clock, so to speak.*

"Local knowledge" or institutional knowledge related to current operational settings retained and passed on from experienced pilots is equally important as formal classroom training. Local knowledge can assist new pilots' decision-making through real-time information on critical operational elements like runway conditions, frequency "dead spots" with weak radio reception, and localized weather patterns, to name a few.

This information is essential for "go or no-go" decisions and was more frequently referenced than the information learned from formal ground school. The length of time flying in Alaska exposed pilots to various situations and allowed them to slowly build their local knowledge repertoire. Two respondents reflected on this "Local Knowledge" idea.

Quote 1

So much of what we do here, so many of the landings and takeoffs that we do here are really at your own risk. You don't have a government entity saying "Yeah, this landing surface is good; you can take that to the bank." That doesn't exist on a river, that doesn't exist on a pond someplace . . . or in a remote region of Alaska. That's a huge part of aeronautical decision-making. If I've never been there before, I'm going to talk to somebody who has.

Quote 2

I think it's invaluable. It sets you up to be ahead of the game. You know what to expect, you know what's coming, and you know what to look out for versus just blasting off and not knowing and then getting caught off guard. So I think, especially operating in Alaska, you need to have as many tricks in your back pocket as possible and hopefully you don't run out of them in the process.

INFRASTRUCTURE LIMITATIONS

TABLE 6.4 Theme 4: Infrastructure Limitations

CODE	*DESCRIPTION*	*COVERAGE*
Lack of services	IFR infrastructure, navigation aids, communication with ATC, NOTAMs, reliability of weather reporting, weather cameras, in-flight services	87.5%
Flying intensity	Extreme weather, varied terrain, difficulty with communication, aircraft conditions, etc.	87.5%
Self-reliance	Necessity to gather own information and make decisions without the help of dispatch, company, etc. Formulate your own plans, survive in harsh environments	75%

Part of the challenge of flying in remote settings is the availability of reliable aviation infrastructure. Infrastructure limitations faced by pilots are shown in Table 6.4. Navigational aids, instrument approaches, weather reporting stations, airport condition reports, radio communication, and radar coverage with air traffic control are all elements of an aviation infrastructure that may be limited in a remote setting.

A lack of any of these elements makes aeronautical decision-making more difficult, predisposing pilots to a higher safety risk of operational hazards. The unavailability or limited functionality of these critical infrastructures constrain operators, who sometimes have to dispatch aircraft under IFR requirements while juggling the legality of such flights under Federal Aviation regulations.

For example, an airport might have an instrument approach procedure for available runways but no weather reporting system on the airport. Suppose weather conditions at the time of arrival require the pilot to fly an instrument approach to land. In that case, the crew must be provided with or have access to an accurate and timely weather report provided by either the on-site weather station or a neighboring airport (as published on the approach procedure). If this is not provided, there are some workarounds.

Pilots fly an approach to a proximal airport other than the destination. Once clear of the weather (or beneath it), they can cancel their IFR flight plan with air traffic control and continue VFR (scud run) underneath the cloud layer by following visual landmarks to their destination. While this is technically legal as long as a pilot has the required clearances from clouds, it sometimes incentivizes unsafe behaviors such as plan continuation errors and limits options should something go amiss while attempting to land at the destination airport. An interview participant described another not-so-savory technique:

> *Basically, punch direct-enter-enter into the* GPS, *fly the synthetic vision on their screen, and build themselves an approach as they fly the video game all the way down to the runway.*

The nature of flying in remote settings means that many routes and destinations do not have weather, instrument approach procedures, or instrument runways. Frequently, the destinations are dirt strips or off-airport altogether. Navigation around mountainous terrain without access to reliable weather updates necessitates a willingness for pilots to alter their plan or return to their departure location before they get lost or, worse, are involved in a CFIT accident. Let us look at an example from a participant who was doing some flying for a remote lodge in Alaska:

> *I was going to follow a certain river to the Lodge. I was hearing reports of [the weather] being better past [redacted] Lodge. I don't remember how far I was from it and [the visibility] just kept getting worse and worse and I'm talking to these guys who are on the other side, and they say it's perfectly*

> *good. It got down to, we'll just say really low. And it was just after that I was a little bit farther and I saw this 185 on amphibs emerge out of this cloud right in front of me, right above me. I was like, "50 to 75 ft different that could have been the end of it. We're turning around right now." And that was an eye opener. But you know it's one of those things where your mind wants you to think that you're fine, you'll pass through it, you're talking to guys on the other side of it. Then reality hits where it's like, "We're right here, we're right now, this is the situation." It doesn't matter what it is 5 mi ahead; if you can't make it another mile, none of it matters.*

It sounds like a few factors were at play. Poor weather, no infrastructure, external pressures from other pilots, and the Lodge saying that the weather was good on the other side . . . all influenced this pilot's decision to continue flying in a situation where a mid-air collision was almost imminent, and CFIT was certainly a possibility.

There were varying opinions from our participants on whether there had been improvement regarding regulatory violations to accomplish flights. Three of them stated that sometimes the only two options were to bend the rules or not fly at all; for them, this attitude had remained relatively constant throughout their careers.

Interestingly, however, the three participants with the most experience (20, 15, and 10 years) flying in Alaska noticed a positive trend away from being backed into this decision-making corner. They noted that more regulatory oversight and improvement to the IFR infrastructure had been apparent throughout their careers and contributed to this positive trend. Despite this, it was emphasized that there is still a need for further aviation infrastructure development.

Air traffic management (ATM) infrastructural limitations invariably constrain service coverage and adversely impact IFR flying in Alaska. Widespread areas without radar coverage or control exist across the state. Furthermore, there are places where contacting air traffic control via very high frequency (VHF) radio communication is difficult, if not impossible.

Frustratingly, these locations often coincide with areas without cellular phone coverage. This is just one example of an operational obstacle to flying in Alaska. Beyond that, pilots need to think ahead when flying to destinations without services they may find at a larger airport. A respondent had this to say:

> *From learning how to land, to getting in and out legally with the IFR system's deficiencies. You can't just file IFR. What do you do when you get to minimums and can't contact Center? Or your airplane is iced up, and the boots don't work. If you land and are covered in ice, what are you going to do? Chip it off? There is no equipment.*

Operators depend on timely and accurate information to make safe decisions about where to land. In the absence of this information, pilots are forced to make decisions in the blind. Unreliable information regarding runway conditions is a common theme in the bush.

In smaller communities, local tradespeople are hired to do runway maintenance and inspections. Inappropriate or nonexistent equipment for snow clearing, grating, and runway servicing often makes runways unusable, which means pilots go to great lengths to acquire information regarding runway conditions. A statement from a respondent highlights this point:

> *I would say probably half of the places we go it's pretty unreliable. There are only some airports up here that are attended at all hours of the night and DOT [department of transportation] is out there plowing or looking at the runway conditions. So sometimes we're lucky that they go out and do that and let us know if there's birds on the runway, or snow on the runway, or ice, and what it looks like to them before we land. There are a lot of places that we go that we don't get that information. And at least a lot of the people that accept our cargo, and some of the radio people . . . she can't tell us exactly what the conditions are, but she can tell us if she has had to scrape off her car window because it was frosty or if she had a foot of snow on her car or not.*

Challenging weather, however, is only one part of what makes flying in Alaska difficult; geography and weather patterns augment the already tricky aspects of flying and navigating with limited infrastructure. "Extreme" is a great adjective to describe operating in the state—preparing to contend with everything in an extreme manner is part of what it takes to fly in remote settings. One interview participant summed up their flying experiences:

> *I think as a pilot, you should already be ready for anything to happen. It's almost more extreme here. You have to be ready for everything here if that makes sense. How many things could go wrong that could kill you, and it happens very often. I think a lot more often than it does happen down south. It seems like there's an endless list of things that I need to be concerned about up here. I'm still learning, and I'm still adding things to that list every day. We have so many things that could cause problems, like the weather and terrain, you know, mud on the runway, landing out in the bush, in the villages.*

Therefore, self-reliance is a critical skill. We have talked about the limited infrastructure, the lack of reliable information, and communication difficulties. Couple this with a tendency for operations to take place in a single pilot scenario, and the element of being out alone in the backcountry starts to hit home.

Nevertheless, how can some of these issues be mitigated? An important factor is information sharing. Local knowledge is like gold to a bush pilot. It provides insights into the intricacies of specific locales and best practices for operation. Whether this is shared through mentorship or camaraderie-like relationships, incremental exposure to new environments, or hangar talk, disseminating local knowledge is one more way for pilots to develop self-reliance and foster a good decision-making toolkit.

The intensity of flying in Alaska requires pilots to be extra mindful of several operational factors that pilots in the lower 48 contiguous states of the United States may take for granted. Access to a controlled airfield, basic ground handling facilities, and lounge/reception facilities for passengers and crew can be nonexistent in remote parts of Alaska. A pilot must consider these factors in the flight planning process. Here is some insight into what goes through a bush pilot's mind when prepping for a flight:

> *There are extremes even before you get in the air you're having to think about. What do you have on you? Where are you going to go, who are you going to talk to? I'd never thought before I came up to Alaska that I would be shooting approaches to airports without talking to people. I had no concept of that. It truly is a unique place because the pilot has to think through everything in their day. Not just the flight and getting on the ground but after I get on the ground. How do I keep us warm? How do I keep us going? How do I get ahold of somebody to cancel the flight plan and tell them we made it? All sorts of strange things.*

ORGANIZATIONAL FACTORS

TABLE 6.5 Theme 5: Organizational Factors

CODE	*DESCRIPTION*	*COVERAGE*
Management	Interactions with management, attitudes surrounding operating principles, managerial culture, management-employee relationships	75%
Competition	Competition between companies or among employees	37.5%
Efficiency	Completing tasks on time, meeting customer demands, scheduling demands, and duty days	100%

The perceptions of the pilots we interviewed were overwhelmingly positive when asked about the leadership at their company. Great! This group of pilots felt management was generally easy to reach out to and supported the pilots. They "had their back," so to speak. We can go home now!

Unfortunately, sometimes, good relationships do not translate into positive feelings about operations. Table 6.5 summarizes the organizational factors that affect pilot decision-making. Aviation is a business; despite emphasizing safety, planes must fly to make money. While emphasizing the need to get the job done, management would occasionally assign flights to pilots more likely to bend the rules. This indicated a gap in the attitudes of both management personnel and the pilot group. The philosophies toward decision-making were not standard between personnel in each group. A pilot reiterated the following:

Unfortunately, management knows some guys are willing to do anything to get in, no matter the weather. If I go "missed" on the approach and return, they'll just send someone else.

Other pilots emphasized a need for management to play a more active role in daily operations. Managers were out of touch with the type and amount of work employees were asked to accomplish.

The high physical demand of loading/unloading cargo, fatigue accompanying multi-leg days, and complex weather challenges could be emphasized if management occasionally rode along with their pilots. This would also foster stronger relationships between management and the pilot groups, potentially encouraging pilots to stay longer (therefore avoiding a high turnover rate—a win-win for the company and pilot). To buttress this point, a pilot noted:

I think we need to start saying, "I'm sorry, I can't do that." If people flew with us, they would say, "Wow." If they had to do it, they would think differently. If they compensated people, they would stay longer. People come here to get experience and leave.

In addition to the disconnect between the realities of flight operational challenges between pilots and management, there appears to be a disconnect between other personnel (like outstation representatives, customer service, and even passengers) on what is required for flight safety. All of these people produce external pressures on pilots to complete flights, even if the conditions do not allow for it and it was noted by one of our interview respondents:

Sometimes, communities expect too much out of pilots regarding why they are not flying or getting in. There are base managers who are not pilots who expect you to work miracles.

Another pilot highlighted how this affects their decision-making:

Thinking, "Hey, this is Alaska; we're still trying to get flights done." That has pushed me more out of safety than anything else. "Know the rules, but know we're trying to get it done." That has happened, pushing you to do something. I've seen it a lot.

Despite pressures from company personnel to complete flying, one participant highlighted the dichotomy between efficiency and safety and discussed how pilots are ultimately in control of finding that balance.

The number one priority is to get the job done, but not at the compromise of safety. In 135, your job is to get the job done, and the responsibility of making all the decisions falls on the pilot. Ultimately, you can say no.

Flying in Alaska requires creativity. If your default is 'no' when asked to get a job done, you won't be successful up here. The default answer needs to be, 'Let me try to figure out a solution.'

Company reputation is important. However, external time-related pressures' effects on creating a culture where pilots fly at all costs lead to a poor safety culture at best and accidents at worst. This goes hand in hand with competition between companies. If one company is flying, there is more pressure for the other local operators to be flying, too.

The Promech case study discussed in Chapter 4 alludes to this—some companies elected to stop flying, and others chose to continue, even in the deteriorating weather conditions. While the pressure exists, it was noted that individual pilots at other companies are keen to share advice and information about weather and runway conditions. In other words, competition does not withhold safety-critical information. To sum up the relationship between competitors, a pilot stated:

For the most part, everyone is looking out for each other.

REGULATORY OVERSIGHT

TABLE 6.6 Theme 6: Regulatory Oversight

CODE	*DESCRIPTION*	*COVERAGE*
Legality	Regulations, rule-bending, working within the 135 regulatory sphere	87.5%
Oversight	FAA oversight, safety reporting	50%

We have heard about the concept of safe and legal decisions a few times. We summarize two regulatory factors in Table 6.6. The inadequate IFR infrastructure (e.g., navigational aids, weather reporting systems, radar coverage) limits operational capabilities and impacts safety when pilots are forced to make decisions that may be operationally efficient but are not necessarily legal.

Remote setting pilots in Alaska sometimes juggle safety, efficiency, and legality options. This requires creativity on the part of the pilot to accomplish all, and sometimes, there are circumstances where the pilot needs to choose one option. The legal option may not be the safest, most efficient, or most reliable.

Pilots who have flown in Alaska for a lengthy period note that there has been improvement in this area—20 years ago, safety-critical flight decisions that were legally questionable were routine. Today, they are less frequent.

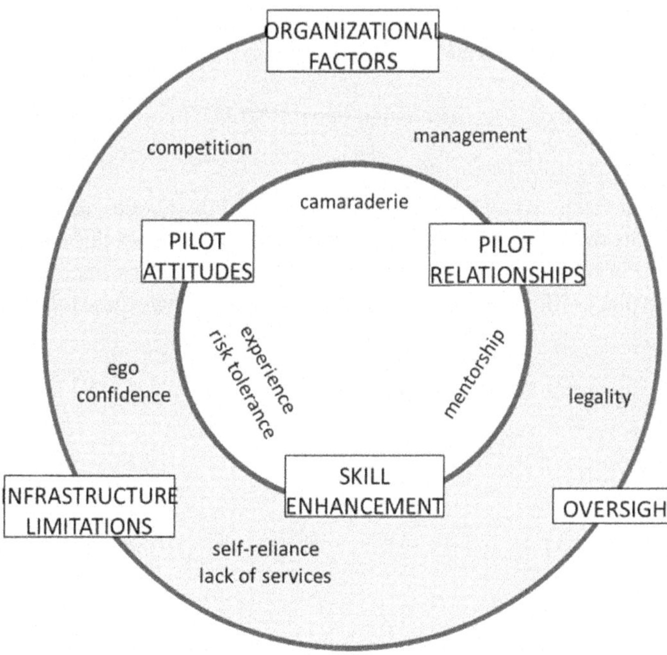

FIGURE 6.1 The relationship between external themes (outside ring) and internal themes (inside ring)

Source: Atkins, 2021.

Despite this improvement, even the newest pilots noted the occasional blurred line between task completion and legality:

> *Sometimes you need to make decisions that are illegal to save you. That happens a lot in Part* 135. *I think any pilot would choose that. Sometimes legal doesn't mean safe.*

Company culture is important in the "safe versus legal" conversation. Safety reporting is difficult when everyone knows everyone at smaller companies. Anonymity is a luxury reserved for large pilot groups. Because of this, pilots are less likely to share information with management through their safety reporting systems.

Pilots talking among themselves and sharing local knowledge and stories is an effective (albeit informal) way to create a good safety culture among the pilot group. The best ways to share and disseminate information were to encourage openness among the pilot group, share relevant information from management, and establish a non-punitive, anonymous reporting system. A respondent put it into context, noting:

> *People are laissez-faire and honest about what they share in hangar talk versus on the record.*

Having discussed all the identified themes and their codes, as well as their implications for policies and practices in Alaska, we wrap up this chapter with a diagram of the complex interrelationships between factors that make flying in remote places like Alaska challenging. Figure 6.1 shows these relationships.

Push the Throttle, Let's Fly Safely in Alaska's Remote Settings 7

Concluding Perspectives and Reflections

After a tour into the world of Part 135 flying in Alaska, we will end with a recap of the various factors that make decision-making challenging in the state. While Alaska's geography is unique, the challenges associated with flying in such a place are not foreign to other remote settings globally. Alaska provided us with a setting to dive deep into the perspectives of its pilots and learn from their experiences.

DOI: 10.1201/9781032678863-7

ENHANCING AERONAUTICAL DECISION-MAKING IN ALASKAN AVIATION OPERATIONS USING INSIDER PERSPECTIVES

Understanding the challenges of flying in remote settings was crucial to see how Alaskan pilots make decisions before and during flight. These significant challenges, many unique to remote settings, all build on each other and amplify the complexity of flying in these regions. By examining the daily obstacles and mitigation strategies pilots in Alaska employ, we gained insight into aeronautical decision-making and human factors elements of flying in such a remote region.

Insider perspectives were the cornerstone of our research. While looking at theoretical perspectives, case studies, accident data, and regulatory updates was important, talking to pilots who actually fly in Alaska made our research truly valuable. Their experiences gave us an honest glimpse into what it takes to operate in one of the wildest places in the United States and overcome the significant hurdles to doing their jobs safely.

As we have discussed, numerous environmental factors make flight operations in Alaska challenging. The expansive geography, varied terrain, and regional differences contribute to these factors. It is common for pilots to fly through many different regions during one flight or duty period. Each region may have its own localized challenges, further complicating pilots' decision-making.

Weather, which can change quickly, often limits pilots from completing flights. There are extreme swings in weather patterns between seasons in Alaska that pilots are forced to contend with. Weather specifically challenging in Alaska are white-out conditions, large expenses of instrument meteorological conditions, long darkness in winter months, and extreme cold temperatures. NTSB accident data shows how low ceilings and visibility were more prevalent in serious and fatal accidents than in accidents with no injuries. Insider perspectives confirm that understanding how to read the weather—both from weather products and during flight—can make or break the outcome of flights.

We also discovered that airport facilities, information, and conditions make flight operations in Alaska challenging. Most airports used by our study respondents are not required to adhere to CFR Part 139 certification standards. Part 139 defines requirements for airports that serve scheduled passenger-serving operations of more than nine seats or unscheduled operations configured with 31 or more seats. These regulations set standards for snow and ice removal, airport rescue and firefighting, airport condition reporting, airport signs, markings, and lights. Alaskan airports are exempt when they only serve small air carriers with 30 or fewer seats (FAA, 2024).

Sometimes, information on runway status at Alaskan airports is difficult to obtain and sometimes unavailable. In instances where pilots are landing off-airport, they rely on local knowledge and their abilities to adequately read environmental conditions before landing. In the event of an aircraft mechanical malfunction, maintenance services are difficult to find outside large city centers. Interestingly, many airports pilots operated to or from lacked basic support services such as fuel, pilot rest facilities, or shelter, highlighting the remoteness of many commercial operations.

This lack of basic service support for flight operations at airports indicated that the remoteness of Alaska contributed to specific job-related concerns for pilots and confirmed findings of previous studies, which suggested that reliable flight-related information may be difficult to find in Alaska (FAA, 2021). Participants intimated that initiatives such as the Weather Camera Project and Capstone Program have contributed to decreasing the positive trends in the rate of accidents and incidents in Alaska. They valued these programs for their usefulness in flight planning and training, respectively. However, the pilots emphasized that accessing reliable flight-related information is still a large hurdle in their decision-making.

Due to the varied nature of Part 135 operations, different types of aircraft equipment are utilized frequently. Aircraft are often modified with skis, floats, or larger tires and landing gear to withstand off-airport landings. These operations require additional training and skill sets to complete flights safely. Pilots are often drawn to the job due to the varied nature of operations and the opportunity to do this unique type of flying. There is no published information regarding landing zone conditions, weather, or services in off-airport operations since the pilots work outside of any airport environment.

Pilots work within a limited IFR system operationally. Navigational aids, weather reporting systems, and radar and communication coverage gaps make operating in the IFR system challenging. Pilots rely on these resources to safely fly IFR across Alaska's diverse geographical regions. Pilots often fly VFR to circumvent limitations with weather reporting, IFR navigational aids, and approaches despite marginal weather conditions.

We discovered that organizational factors impact decision-making and have contributed to accidents in Alaska. Our study findings suggest that sometimes pressures from management—like the need to make money, maintain reputation, and be efficient—influence pilots (Bearman et al., 2009). We also noted that in addition to management, outstation representatives and passengers are often "pushy" and explicitly encourage pilots to make flights in marginal conditions. This could be due to a general misunderstanding of the requirements to safely and legally complete flights. After all, the training for these personnel and passengers is not nearly as expansive as that for pilots.

Despite the generally strong interpersonal relationships between pilots, our findings show that managerial relationships need improvement. This could be due to a disconnect between management and the pilot group on the reality of day-to-day operations. A positive relationship between pilots and management allows pilots to make decisions in the best interest of safety. Pilots are more likely to turn down flights in which they do not see a successful outcome when management will back them. The influence of theoretical concepts like *Goal Seduction* may be lessened when pilots have more support from management to delay or cancel flights.

Some of the mitigating strategies suggested by interview participants, such as management's "ride-along" to observe the working environment of the pilot group, may help close this gap and create better employee-employer relationships. Despite this, we observed that some pilots are susceptible to "foot-in-the-door" persuasion as management occasionally selects specific pilots to make flights in marginal conditions. Creating a work culture where pilots feel empowered to turn down a flight helps ensure a proactive safety culture within an organization.

We discovered that psychosocial factors like camaraderie can positively influence decision-making and contribute to a safety-centered culture. Many companies in Alaska have small pilot groups or pilot bases where relationships between crew members develop organically. In these close-knit environments, it is natural that relationships influence operations. Theories of modeled behavior—like those from social cognitive theory—help explain how positive and negative influences influence pilot decision-making.

A stronger sense of camaraderie can lead to organizational change. Our participants agreed emphatically that these relationships are generally positive, and camaraderie and mentorship help instill a safety-positive attitude. Though some previous studies suggested that negative peer pressure was more influential on decision-making than camaraderie, our participants felt they learned more from positive influences throughout their careers (Bearman et al., 2009; Michalski & Bearman, 2014).

Mentorship was one of the most important factors when determining which group construct influences pilots. Mentorship is a significant part of the Part 135 pilot culture in Alaska and may be successful due to its informal implementation. By creating an environment where pilots learn from one another and use the advice of others, safety culture may improve and lead to reduced incidents and accidents in the state.

The natural camaraderie among pilots with similar positive safety mindsets helps instill a safety-conscious attitude in the pilot group. There are outliers within any group of people, and the general perception is that pilots who prioritize safety earn the most respect. Another notable finding was that ego was not an admirable trait, especially when it influences an organization's safety. These characteristics were becoming less prevalent in Part 135 operations in Alaska.

We found that operational flying experience in Alaska was one of the most influential qualities when determining a pilot's manual flying skill and risk tolerance. Our pilot sample also viewed good experience as paramount to quality decision-making. Furthermore, biological age had minimal impact on decision-making compared to operational flight experiences. This observation could be due to exposure to operational and environmental challenges like weather, terrain, and mechanical malfunctions, which may only be gathered from first-hand experience.

Emphasizing the free flow of flight-related information using strategies like hangar-talk sessions and safety report dissemination may be particularly valuable for less-experienced pilots. This is notable since first-hand experience plays a vital role in shaping decision-making, especially considering the strengthened effect of observing behavior with positive outcomes (Wood & Bandura, 1989).

The perception of safety culture among the Part 135 pilot sample in Alaska was generally positive. Our pilot sample was willing to share their operational experiences for the benefit of the larger pilot group. Participants from small companies exposed challenges in implementing truly anonymous reporting systems. One participant with experience in management suggests it is difficult to have true anonymity when the pilot group is close-knit and small. Participants in other companies describe how they still use simple and non-online methods (pen and paper in a drop-box) to gather safety reports and how these methods are not taken seriously.

WHAT IS NEXT FOR AVIATION IN ALASKA?

Though the accident and incident rates in Alaska are slowly decreasing, the industry seems to flinch each time there is a fatal accident. There have been some changes in recent years within Alaska—the Medallion Foundation closed its doors, the NTSB continues recommending changes to Part 135 flying, and the FAA released its final report on aviation safety in Alaska. Positives and negatives aside, significant work still needs to be done to reduce the accident rates further. This will take collective, intentional practice by pilots, organizational management, and the regulator. Our pilot participants noted some tangible further steps.

The Aviation Safety Action Program (ASAP), which uses zero-fault incident reports to bolster safety programs, has been introduced into some operations. Despite being voluntary, it is one of the initiatives by the FAA to improve safety reporting culture (FAA, 2021). The general perception of these programs was that they could be effective when implemented among Part 135 operators in Alaska. However, participants reiterated the need for assurances that they are non-punitive and will be treated anonymously.

Regarding safety and operational information sharing, we inferred that peer-to-peer sharing of vital information was more relevant to participants' day-to-day jobs than information from the company's top management personnel. This is an interesting observation—the importance of casual mentorship relationships and camaraderie among these participants was substantial. Hangar talks and sharing local knowledge were more valuable for participants' decision-making than company-sponsored materials. This helps point to the disconnect between top management and the pilot group, which must be addressed to further develop a proactive safety culture among Alaskan pilots.

GLOBAL IMPLICATIONS AND INTERNATIONAL PERSPECTIVE

While our research explicitly focused on pilots in Alaska, the results from our study are applicable globally. The hazards and barriers to effective decision-making certainly do not impact only Alaskan pilots. Other studies with a global focus have suggested similar findings regarding the organizational, operational, geographical, and infrastructure challenges that make aeronautical decision-making difficult. Using the perspectives of participants from our study to bolster safety in Alaska could be a good primer for other regions to learn from.

The perspectives of our participants should be leveraged to enhance safety outside of Alaska. Their insights are a valuable resource for aviation communities and operators globally. The multifaceted challenges faced by pilots in remote settings have been echoed through previous research—this detailed study bolsters the understanding of the remote environment and the specific human factors elements that are part of operating within it.

Despite technological advances in air transportation, some remote areas in the world may never have adequate aviation infrastructure, such as navigational and surveillance capabilities, paved aerodromes with handling services, and air traffic control. These may be due to the cost-benefit of significant capital investment due to low passenger and cargo throughputs at these airfields. Other salient environmental factors that stifle investments to improve aviation infrastructure may be topography and weather.

The communities in those areas will still need air transport services such as resupply of essential commodities, medical evacuation, and emergency response during natural disasters. That is why both local and national-level governmental entities must consider it a social contract and national security initiative to invest adequate resources to ensure continuous air serviceability of these remote airstrips. To conclude this portion of global perspectives, Kwasi recalls two eventful flights in the remote parts of Ghana and how the variables are similar to Alaska.

Close Shaves in Northern Ghana

Between 2002 and 2008, I regularly flew the BN-2T Islander and Fokker to remote airstrips such as Navrongo, Manso Nkran, and Wa in Ghana, West Africa. It was always gratifying to see the joy on the faces of the local communities when we carried out medical evacuations of seriously injured accident victims, sent in vaccines for an outbreak of anthrax, and transported government officials to these locations, to name a few of the tasks we undertook. The flights to these airstrips were always challenging, with basically no navigational aids, fuel services, or rescue fire service cover.

These airstrips in northern Ghana could become water-logged and unusable during the rainy seasons or extremely dusty during the Harmattan season, obscuring landing visibility. Even some of the airports in Ghana had infrastructural challenges during this period when I was a pilot conducting these remote-setting operations. I recall doing night flights to airports such as Takoradi and Tamale, which did not have runway lights until 2007. We used lighted "goose lanterns," which was daunting. These goose lanterns were aluminum canisters with spouts (like a gooseneck) filled with either kerosene or Jet-A1 fuel. A wick inside was lit to provide light.

I vividly remember a close shave with CFIT on a flight to Tamale, in northern Ghana, which can have very poor visibility during the Harmattan season from November to February every year. We had been doing some logistics and supply flights to Takoradi, Kumasi, and Sunyani Airports and were returning back to Tamale. We were on the last leg and quite fatigued. After an uneventful cruise, we were cleared for a straight-in VOR approach to Runway 05 in

Tamale around 10:45 PM UTC. The visibility was about three miles in haze due to the Harmattan. These are the intricacies of flying in resource-constraint settings where non-precision approaches at night become the order of the day.

For some inexplicable reason, someone decided to park their vehicle about a mile to the runway threshold and put their bright headlights facing our approach direction as if to guide us while we were on approach to land. That affected our ability to see the lighted "goose lantern" in an environment around the airport with minimal terrain; we experienced a "black-hole "illusion. This illusion can occur during a night approach toward an airport environment with few ground features and lights. You perceive that you are higher than you actually are. Pilots tend to fly lower approaches into these kinds of airports, hence the name "black hole effect." The darkness sucks you in, and if you are not careful, it can cause you to crash short of the runway.

We were inadvertently flying toward the light from the parked vehicle off the threshold until around 500 feet when we selected full flaps for landing. We realized that we were low and about a mile from the threshold of runway 05. The Ground Proximity Warning Systems (GPWS) alerted us. We initiated a go-around, repositioned, and landed uneventfully but shaken to our core when we reflected on our close shave with CFIT in a post-flight debrief.

My second recollection was a medical evacuation flight from Navrongo Airstrip (ICAO: DGLN), which serves Paga and Navrongo, towns in the Kassena-Nankana District of the Upper East Region of Ghana to Accra, the capital of Ghana. The airstrip has an unpaved surface and is designated 14/32 with an elevation of 678 ft above mean sea level. The airstrip was about $4{,}966\,ft\ (1{,}514\,m)$in length. Anytime we flew to Navrongo, we had a procedure to fly over the field at about 1000 feet above the terrain to inspect and ensure the field was safe and secure before initiating an approach to land. We did exactly that, and the initial flyover alerted the medical team with the patient to be evacuated of our arrival.

On the low pass, we did not see any safety issues on the field, and we positioned downwind and a base turn that sometimes meanders into the border with Burkina Faso, Ghana's neighbor to the North. We made a short-field approach since some power pylons at the edge of the airstrip posed a hazard for any approach, and that required clearing before a steep descent for landing. Landing in Navrongo was always a test of airmanship and flying skills. Normally, newer pilots in my organization had to do a couple of airstrip landings under the supervision of a senior pilot or training captain before being released to fly there unsupervised.

After flaring and touching down on the airstrip, my co-pilot called for a "wave off" or go-around/aborted landing as I saw what was straight ahead. Without hesitation, I added maximum power, and we initiated a "wave off," which is basically executing a takeoff just after landing for a transport category aircraft like the Fokker F-27. Why did we "wave off" or execute a go-around? A herd of stray cattle suddenly meandered their way onto the latter end of the airstrip, causing a potential collision hazard even if we attempted maximum braking to stop short of that area. This occurred because the strip was unsecured and uncontrolled, and wildlife crossing was always an inherent risk.

We climbed up to 1,000 feet and made another low pass over the field to alert anyone around the airstrip that we intended to land. By now, the medical team had realized what had happened, and they rallied to get the herdsman to muster the herd away to a safe distance from the airstrip with the assistance of some local farmers and curious bystanders at the airstrip. We repositioned, still wary and on the lookout, and did a second landing without any issues.

One of the challenges of flying into these remote airstrips in Ghana was that we had to create our own GPS approaches because there were no approved aerodrome information publications (AIP). Using hand-held GPS, we took coordinates of various airstrips on the ground and noted high terrains and obstacles within a 25-mile radius of the fields. We also adhered strictly to minimum obstacle clearance heights, and most flights were in day VFR conditions.

The lesson here is that in remote settings such as Navrongo in Ghana, the outback of Australia, tropical forests in the Democratic Republic of Congo, Kalahari deserts in Namibia, mountainous regions in Papua New Guinea, or "bush country" in Alaska, effective decision-making starts with comprehensive pre-flight planning and operational risk assessments.

FIGURE 7.1 A Ghana Airforce Fokker-F27 Mark 400 flown by the author (Kwasi) to various remote airstrips in Ghana

Source: Kwasi Adjekum.

FIGURE 7.2 A red and white single-engine DHC-2 Beaver aircraft on floats sails in to a rocky beach on a calm emerald-green lake surrounded by mountains on Kodiak Island, Alaska.

Source: Dana Atkins.

Pilots flying in remote settings require a continuous decision-making process when inflight to assist in terminating or proceeding with a flight based on the variability of the weather and airstrip surface conditions. Decision-making in these settings is primed by expert knowledge of the environment, experiences gained over time flying in the environment, and an understanding of safety risk assessments. It also requires professionalism and knowing the limits of an individual's expertise and human performance capability.

CONCLUDING REMARKS

So, after interacting with our participants and digesting their insights, what did we learn? We remind the reader that in Alaska, flight operations support daily life. They are critical to the Alaskan economy. They provide tourism and recreational activities. Developing and sustaining a cadre of Part 135 pilots who can operate safely and efficiently within the Alaskan aviation environment is essential. Alaska has long been portrayed as a hotspot for aviation hazards. Our study brings to light the remarkable efforts to improve aviation safety from the viewpoints of key stakeholders operating in the far reaches of the state. It is one thing to interpret from afar; it is something completely different to hear (or have!) firsthand experience. The perspectives of pilots working in the environment carry significant weight and should be given attention when discussing aviation safety improvements.

Creating mentorship programs in Part 135 operations in Alaska that pair experienced and safety-conscious pilots with new pilots can strengthen camaraderie and safety culture within companies. Tapping into the operational experience of seasoned pilots can help shape policy as it is implemented. The finding that safety-critical information is most effective in a casual setting allows operators to adopt informal and formal methods when sharing critical information. This blend of informal and formal strategies can foster a robust safety culture. The use of social media platforms and hangar talks could be beneficial for operators. Emphasizing local knowledge could guide FAA initiatives to gather and disseminate this information through informal but widespread and accessible means.

Finally, a push for continuing aviation infrastructure developments from local and Federal sources that address shortcomings in weather reporting, navigational aids, and radar coverage will help give pilots options and provide information that may ultimately prevent accidents. The FAA Alaska Aviation Safety Initiative Final Report, published in October 2021, underscores the need to improve this infrastructure (FAA, 2021). Despite the monetary challenges, continuing improvement is the most forthright change to improve aviation safety in Alaska.

Chapter References

CHAPTER 1

Alaska Aviation System Plan (2020). *Alaska Weather Equipment Needs Summary.* State of Alaska Department of Transportation. https://www.alaskaasp.com/media/1897/ak_weather_bundle_update.pdf

Anderson, J. (2000). *Arctic Bush Pilot: From Navy Combat to Flying Alaska's Northern Wilderness: A Memoir as told to Jim Rearden.* Kenmore, WA: Epicenter Press.

AOPA (2024). *Aircraft Fact Sheet: Cessna 180.* Retrieved (05/07/2024). www.aopa.org/go-fly/aircraft-and-ownership/aircraft-fact-sheets/cessna-180

Conway, G. A., Hill, A., Martin, S., Mode, N. A., Berman, M. D., Bensyl, D. M., Manwaring, J. C., & Moran, K. A. (2004). Alaska air carrier operator and pilot safety practices and attitudes: A statewide survey. *Aviation, Space, and Environmental Medicine, 75*(11), 984–991.

Federal Aviation Administration. (2004). *Seaplane, Skiplane, and Float/Ski Equipped Helicopter Operations Handbook.* U.S. Department of Transportation. www.faa.gov/regulations_policies/handbooks_manuals/aviation/seaplane_handbook/

Federal Aviation Administration. (2018). *Instrument Rating – Airplane Airman Certification Standards (FAA-S-ACS-8B).* www.faa.gov/sites/faa.gov/files/training_testing/testing/acs/instrument_rating_acs_change_1.pdf

Federal Aviation Administration. (2021a, April 23). *FAA Alaska Aviation Safety Initiative Interim Report.* www.faa.gov/sites/faa.gov/files/2021-09/FAASI_Interim_Report.pdf

Federal Aviation Administration. (2021b). *FAA Alaska Aviation Safety Initiative FY21 Final Report.* www.faa.gov/sites/faa.gov/files/2021-10/FAA%20Alaska%20Safety%20Initiative%20Report%20-%20final.pdf

Federal Aviation Administration. (2023). *General Information. 14 CFR Part 135 Operations.* www.faa.gov/licenses_certificates/airline_certification/135_certification/general_info

Federal Aviation Administration. (2024, March 21). *Chapter 2: Aeronautical Information Manual. Section 2: Air Navigation.* Washington, DC. www.faa.gov/air_traffic/publications/atpubs/aim_html/chap1_section_2.html

Foster, J. A. (1990). The bush pilots: A pictorial history of a Canadian phenomenon. *Canadian Book Review Annual Online.* Retrieved (2/5/2024). https://cbra.library.utoronto.ca/items/show/10566

Grabowski, M., Rizzo, R., & Graig, T. (2016). Data challenges in dynamic, large-scale resource allocation in remote regions. *Safety Science, 87,* 76–86. https://doi.org/10.1016/j.ssci.2016.03.021

Horn, C., Gifford, S. M., & Ting, C. Y. P. (2021). Informal, essential and embedded: Transport strategies in remote Sarawak. *Journal of Transport Geography, 96,* 103181. https://doi.org/10.1016/j.jtrangeo.2021.103181

International Civil Aviation Organization. (2011). *Aerodrome Meteorological Observation and Forecast Study Group (AMOFSG), 9th Meeting.* Montreal, Canada. www.icao.int/safety/meteorology/amofsg/AMOFSG%20Meeting%20Material/SN%2021%20Binder1.pdf

Kuklina, M., Filippova, V., Savvinova, A., Fedorova, A., & Itegelova, A. (2022). Influence of transport accessibility on the development of remote areas of residence of the indigenous people of the North. *Transportation Research Procedia, 63,* 41–47. https://doi.org/10.1016/j.trpro.2022.05.005; www.sciencedirect.com/science/article/pii/S2352146522002423

Lee, H. (2023, August 31). *Infrastructure Challenges in the Alaskan Arctic.* Policy Brief, Belfer Center for Science and International Affairs, Harvard Kennedy School. Cambridge, MA. https://www.belfercenter.org/sites/default/files/files/publication/Lee_Infrastructure%20Challenges%20in%20the%20Alaskan%20Arctic%20Brief_FINAL.pdf

Michalski, D. J., & Bearman, C. (2014). Factors affecting the decision making of pilots who fly in Outback Australia. *Safety Science, 68,* 288–293. https://doi.org/10.1016/j.ssci.2014.03.005

National Transportation Safety Board. (n.d.). *Who We Are and What We Do.* National Transportation Safety Board. https://www.ntsb.gov/Pages/home.aspx

National Transportation Safety Board. (2019a). *Alaska Part 135 Flight Operations, Charting a Safer Course.* www.ntsb.gov/news/events/Pages/2019-MWL-RT-AK-part135.aspx

National Transportation Safety Board. (2019b, February 4). *NTSB Issues 2019–2020 Most Wanted List of Transportation Safety Improvements.* www.ntsb.gov/news/press-releases/Pages/mr20190204.aspx

National Transportation Safety Board. (2020). *Safety Recommendation Report: Revise Processes to Implement Safety Enhancements for Alaska Aviation Operations.* www.ntsb.gov/investigations/AccidentReports/Reports/ASR2002.pdf

National Transportation Safety Board. (2023, June 16). *Aviation Investigation Preliminary Report, Shaktoolik, Alaska.* (Report No. ANC23FA042).

Pocock, M. C. C. (2017). *Bush & Mountain Flying: A Comprehensive Guide to Advanced Bush & Mountain Flying Techniques and Procedures* (3rd rev.). Digital Publishing of Florida Inc.

Ringsmuth, K. J. (2015). *Alaska's Skyboys: Cowboy Pilots and the Myth of the Last Frontier.* Seattle, WA: University of Washington Press.

Rodrigue, J.-P. (2024). *The Geography of Transport Systems* (6th ed.). New York: Routledge.

Ross, A. (2015, May 13). The early female aviator who flew across the Atlantic in terrible weather. *Atlas Obscura Magazine.* www.atlasobscura.com/articles/daring-aviator-beryl-markham

Ruffing, K., & Cochrane, D. (2023, March 29). *Three Women in Aviation You May Not Have Heard of.* Smithsonian National Air and Space Museum. https://airandspace.si.edu/stories/editorial/three-women-aviation-you-may-not-have-heard

State of Alaska Division of Homeland Security. (2024). *Emergency Operations Plan.* Retrieved (5/2/2024). State of Alaska EOP.pdf; https://ready.alaska.gov/Documents/Plans/2024%20State%20of%20Alaska%20EOP.pdf

CHAPTER 2

Aircraft Owners and Pilots Association. (2021, January 12). *FAA Pitches Safety Reporting Programs to Part 135 Operators.* Retrieved (05/07/2024). www.aopa.org/news-and-media/all-news/2021/january/12/faa-pitches-safety-reporting-program-to-part-135-operators

Alaska Aviation: Field Hearing before the U.S. Senate Committee on Commerce, Science, and Transportation, 109th Congress. (2005). www.commerce.senate.gov/services/files/6507C9D5-1D57-4FD0-9F88-0D8C8F61B788

Appendices: Alaska Mapping Business Plan [PDF]. (2017, September). Anchorage, Alaska: Alaska Department of Commerce, Community, and Economic Development Division of Community and Regional Affairs. https://www.commerce.alaska.gov/web/dcra/PlanningLandManagement/RiskMAP/AlaskaMappingBusinessPlan.aspx

Bearman, C., Paletz, S. B. F., Orasanu, J., & Brooks, B. P. (2009). Organizational pressures and mitigating strategies in small commercial aviation: Findings from Alaska. *Aviation, Space, and Environmental Medicine, 80*(12), 1055–1058. http://dx.doi.org/10.3357/ASEM.2590.2009

Berman, M., Martin, S., & Hill, A. (2005). *Evaluation of the Alaska Interagency Aviation Safety Initiative.* Institute of Social and Economic Research, University of Alaska. http://hdl.handle.net/11122/12221

CAPA. (2019). *Anchorage Airport: An East-West Growing Hub into Cargo and e-Commerce.* CAPA. centreforaviation.com

Causse, M., Dehais, F., Péran, P., Sabatini, U., & Pastor, J. (2013). The effects of emotion on pilot decision-making: A neuroergonomic approach to aviation safety. *Transportation Research Part C: Emerging Technologies, 33,* 272–281. https://doi.org/10.1016/j.trc.2012.04.005

Cloe, J. H. (2015, February 9). *Wide Open on Top: Alaska and the Cold War.* Alaska Historical Society. https://alaskahistoricalsociety.org/wide-open-on-top-alaska-and-the-cold-war/

Dismukes, R. K., Goldsmith, T. E., & Kochan, J. A. (2015, August). *Effects of Acute Stress on Aircrew Performance: Literature Review and Analysis of Operational Aspects.* Langley, VA: National Aeronautics and Space Administration (NASA)

STI Program. Retrieved (5/7/2024). https://human-factors.arc.nasa.gov/publications/NASA_TM_2015_218930-2.pdf

Federal Aviation Administration (2012). *Chapter 5–31: Instrument Flying Handbook (FAA-H-8083-15B).* https://www.faa.gov/sites/faa.gov/files/regulations_policies/handbooks_manuals/aviation/FAA-H-8083-15B.pdf

Federal Aviation Administration. (2020, March 3). FAA Weather Cameras. www.faa.gov/about/office_org/headquarters_offices/ato/service_units/systemops/fs/alaskan/weather_cams/

Federal Aviation Administration. (2021a, April 23). *FAA Alaska Aviation Safety Initiative Interim Report.* www.faa.gov/sites/faa.gov/files/2021-09/FAASI_Interim_Report.pdf

Federal Aviation Administration. (2021b). *FAA Alaska Aviation Safety Initiative FY21 Final Report.* www.faa.gov/sites/faa.gov/files/2021-10/FAA%20Alaska%20Safety%20Initiative%20Report%20-%20final.pdf

Hollander, Z., & McGee, M. (2019, August 19). Aviation safety auditor medallion foundation to close in September. *Anchorage Daily News.* www.adn.com/alaska-news/aviation/2019/08/18/aviation-safety-auditor-medallion-foundation-to-close-in-september/

Kraus, T. L. (2020). *Civil Aviation Policy in Alaska, 1913–2018.* U.S. Department of Transportation, Federal Aviation Administration. Washington, DC. https://www.faa.gov/sites/faa.gov/files/about/office_org/headquarters_offices/arc/Civil_Aviation_Policy_in_Alaska.pdf

Lai, H., Chen, C., Khoo, L. P., & Zheng, P. (2019). Unstable approach in aviation: Mental model disconnects between pilots and air traffic controllers and interaction conflicts. *Reliability Engineering and System Safety, 185,* 383–391. https://doi.org/10.1016/J.RESS.2019.01.009

Leigh, L., & Rissanen, R. (2010). Recruiting and training better 'bush pilots': A research-based approach. *Journal of Aviation Management and Education,* (1) 1–16. https://aabri.com/manuscripts/09283.pdf

Madhavan, P., & Lacson, F. C. (2006). Psychological factors affecting pilots' decisions to navigate in deteriorating weather. *North American Journal of Psychology; Winter Garden, 8*(1), 47–62.

Michalski, D. J., & Bearman, C. (2014). Factors affecting the decision making of pilots who fly in Outback Australia. *Safety Science, 68,* 288–293. https://doi.org/10.1016/j.ssci.2014.03.005

Mode, N. A., O'Connor, M. B., Conway, G. A., & Hill, R. D. (2012). A multifaceted public health approach to statewide aviation safety. *American Journal of Industrial Medicine, 55*(2), 176–186. https://doi.org/10.1002/ajim.21993

Mondor, C. (2021, March 4). Medallion Foundation post-mortem: The safety program that didn't deliver. *Skies.* https://skiesmag.com/news/medallion-foundation-post-mortem-the-safety-program-that-didnt-deliver/

National Institute for Occupational Safety and Health. (2018, October 22). *Aviation Safety Research Program.* Centers for Disease Control and Prevention. www.cdc.gov/niosh/topics/aviation/

National Transportation Safety Board. (2017a). *Collision with Terrain, Promech Air, Inc. de Havilland DHC-3, N270PA, Ketchikan, AK.* June 25, 2015 (Aircraft Accident Report No. NTSB/AAR-17/02). www.ntsb.gov/investigations/AccidentReports/Reports/AAR1702.pdf

National Transportation Safety Board. (2017b). *Investigative Hearing Flight 3153 Anchorage, Alaska Aug 17, 2017 pt. 1.* [Video]. YouTube. https://www.youtube.com/watch?v=ZrUYwV4oECI&t=1s

National Transportation Safety Board. (2018). *Collision with Terrain, Hageland Aviation Services, Inc. dba Ravn Connect Flight 3153, Cessna 208B, N208SD, Togiak, Alaska.* October 2, 2016 (Report No. NTSB/AAR-18/02). www.ntsb.gov/investigations/AccidentReports/Reports/AAR1802.pdf

National Transportation Safety Board. (2019a). *Alaska Part 135 Flight Operations, Charting a Safer Course.* www.ntsb.gov/news/events/Pages/2019-MWL-RT-AK-part135.aspx

National Transportation Safety Board. (2019b, February 4). *NTSB Issues 2019–2020 Most Wanted List of Transportation Safety Improvements.* www.ntsb.gov/news/press-releases/Pages/mr20190204.aspx

National Transportation Safety Board. (2021, April 15). *Require and Verify the Effectiveness of Safety Management Systems in All Revenue Passenger-Carrying Aviation Operations.* www.ntsb.gov/Advocacy/mwl/Pages/mwl-21-22/mwl-as-01.aspx

National Transportation Safety Board. (2022). *Aviation Investigation Final Report, Ketchikan, Alaska.* August 5, 2021 (Report No. ANC21FA069).

Noort, M. C., Reader, T. W., & Gillespie, A. (2021). Safety voice and safety listening during aviation accidents: Cockpit voice recordings reveal that speaking-up to power is not enough. *Safety Science, 139,* Article 105260. https://doi.org/10.1016/j.ssci.2021.105260

Orasanu, J. M. (2010). Flight crew decision-making. In B. G. Kanki, R. L. Helmreich, & J. Anca (Eds.), *Crew Resource Management* (2nd ed., pp. 147–179). Academic Press. https://doi.org/10.1016/B978-0-12-374946-8.10005-6

Paletz, S. B. F., Bearman, C., Orasanu, J., & Holbrook, J. (2009). Socializing the human factors analysis and classification system: Incorporating social psychological phenomena into a human factors error classification system. *Human Factors, 51*(4), 435–445. https://doi.org/10.1177/0018720809343588

Ringsmuth, K. J. (2014, December). *Bob Reeve: Eastern Alaska's early Air Prospector, 1932–1938.* National Park Service. www.nps.gov/articles/aps-v13-i2-c1.htm

Ringsmuth, K. J. (2015). *Alaska's Skyboys: Cowboy Pilots and the Myth of the Last Frontier.* Seattle, WA: University of Washington Press.

Ringsmuth, K. J. (n.d.). *The Wrangell Mountain Skyboys: Making History Above the Copper Belt.* Alaska Historical Society. https://alaskahistoricalsociety.org/discover-alaska/kids-page/the-wrangell-mountain-skyboys/

Shiner, L. (2013, May 19). Bush pilot hall of fame. *Air and Space Magazine.* www.smithsonianmag.com/air-space-magazine/bush-pilot-hall-of-fame-68839910/

State of Alaska. (n.d.). *Facts about Alaska.* Retrieved October 10, 2021, from https://www.alaska.gov/kids/learn/facts.htm

Strohschneider, S. (2002). Cultural factors in complex decision making. *Online Readings in Psychology and Culture, 4*(1). https://doi.org/10.9707/2307-0919.1030

Transport Canada. (2002). *Human Performance Factors: Pilot Decision Making.* https://tc.canada.ca/sites/default/files/migrated/mod3.pdf

University of Alaska Anchorage. (2005). *The Impact of Capstone Phase 1 Program: Final Report.* www.faa.gov/nextgen/programs/adsb/Archival/media/Phase%201%20Final%20with%20appendices.pdf

CHAPTER 3

Adhikari, P. (2015). Pogreške i nesreće na radnom mjestu [Errors and accidents in the workplaces]. *Sigurnost, 57*(2), 127–137.

Aircraft Accident Investigation Commission, Government of Nepal. (2016). *Final Report on 9N-AHH Twinotter (Viking DHC6-400).* Aircraft Accident Operated by Tara Air Pvt Ltd. Retrieved (5/8/2024). https://reports.aviation-safety.net/2016/20160224-0_DHC6_9N-AHH.pdf

Bandura, A. (1986). *Social Foundations of Thought and Action.* Englewood Cliffs, NJ: Prentice Hall.

Barron, L. G., Carretta, T. R., & Bonto-Kane, M. V. A. (2016). Relations of personality traits to military aviator performance. *Aviation Psychology and Applied Human Factors, 6*(2), 57–67. https://doi.org/10.1027/2192-0923/a000100

Bearman, C., Paletz, S. B., & Orasanu, J. M. (2009). Situational pressures on aviation decision making: Goal seduction and situation aversion. *Aviation, Space, and Environmental Medicine,* 80(6), 556–60. https://doi.org/10.3357/ASEM.2363.2009

Breuer, S., Ortner, T. M., Gruber, F. M., Hofstetter, D., & Scherndl, T. (2023). Aviation and personality: Do measures of personality predict pilot training success? Updated meta-analyses. *Personality and Individual Differences, 202,* 111918. https://doi.org/10.1016/j.paid.2022.111918

Brown, R. (2000). Social identity theory: Past achievements, current problems and future challenges. *European Journal of Social Psychology, 30*(6), 745–778. https://doi.org/10.1002/1099-0992(200011/12)30:6<745::AID-EJSP24>3.0.CO;2-O

Burford, B. (2012). Group processes in medical education: Learning from social identity theory. *Medical Education, 46*(2), 143–152. https://doi.org/10.1111/j.1365-2923.2011.04099.x

Causse, M., Péran, P., Dehais, F., Caravasso, C. F., Zeffiro, T., Sabatini, U., & Pastor, J. (2013). Affective decision making under uncertainty during a plausible aviation task: An fMRI study. *NeuroImage, 71,* 19–29. https://doi.org/10.1016/j.neuroimage.2012.12.060

Clarke, S., & Robertson, I. T. (2005). A meta-analytic review of the big five personality factors and accident involvement in occupational and non-occupational settings. *Journal of Occupational and Organizational Psychology, 78*(3), 355–376. https://doi.org/10.1348/096317905X26183

Dekker, S. (2014). *The Field Guide to Understanding "Human Error"* (3rd ed.). CRC Press. http://dx.doi.org/10.1201/9781317031833

Detwiler, C., Hackworth, C., Holcomb, K., Boquet, A., Pfleiderer, E., Wiegmann, D., & Shappell, S. (2006). *Beneath the Tip of the Iceberg: A Human Factors Analysis of General Aviation Accidents in Alaska versus the Rest of the United States.* (Technical Report No. DOT/FAA/AM-06/7). U.S. Department of Transportation, Federal Aviation Administration, Office of Aerospace Medicine, Civil Aerospace Medical Institute. www.faa.gov/data_research/research/med_humanfacs/oamtechreports/2000s/media/200607.pdf

Detwiler, C., Holcomb, K., Hackworth, C., & Shappell, S. (2008). *Understanding the Human Factors Associated with Visual Flight Rules Flight into Instrument Meteorological Conditions.* (Technical Report No. DOT/FAA/AM-08/12). U.S. Department of Transportation, Federal Aviation Administration, Office of Aerospace Medicine, Civil Aerospace Medical Institute. https://apps.dtic.mil/sti/citations/ADA482973

Dolinski, D. (2012). The nature of the first small request as a decisive factor in the effectiveness of the foot-in-the-door technique. *Applied Psychology: An International Review, 61*(3), 437–453. https://doi.org/10.1111/j.1464-0597.2011.00477.x

Elms, A. K., Gill, H., & Gonzalez-Morales, M. G. (2023). Confidence is key: Collective efficacy, team processes, and team effectiveness. *Small Group Research, 54*(2), 191–218. https://doi.org/10.1177/10464964221104218

English, D., & Branaghan, R. J. (2012). An empirically derived taxonomy of pilot violation behavior. *Safety Science, 50*(2), 199–209. https://doi.org/10.1016/j.ssci.2011.08.009

Federal Aviation Administration. (2023, July 10). *Chapter 2–5: Pilots Handbook of Aeronautical Knowledge (FAA-H-8083-25B).* www.faa.gov/sites/faa.gov/files/04_phak_ch2.pdf

Fitzgibbons, A., Schutte, P., & David, D. (2000). *Pilot Personality Profile using the NEO-PI-R.* https://ntrs.nasa.gov/citations/20000105204

Flippen, A. R., Hornstein, H. A., Siegal, W. E., & Weitzman, E. A. (1996). A comparison of similarity and interdependence as triggers for in-group formation. *Personality and Social Psychology Bulletin, 22,* 882–893.

Ford, J., O'Hare, D., & Henderson, R. (2013). Putting the "we" into teamwork: Effects of priming personal or social identity on flight attendants' perceptions of teamwork and communication. *Human Factors, 55*(3), 499–508. https://doi.org/10.1177/0018720812465311

Freedman, J. L., & Fraser, S. C. (1966). Compliance without pressure: The foot-in-the-door technique. *Journal of Personality and Social Psychology, 4*(2), 195–202. https://doi.org/10.1037/h0023552

Gao, Y., & Kong, S. (2016). Personality types of pilot students: A study of an Australian collegiate aviation program. *International Journal of Aviation, Aeronautics, and Aerospace, 3*(3). https://doi.org/10.15394/ijaaa.2016.1130

Gilbey, A., & Hill, S. (2012). Confirmation bias in general aviation lost procedures. *Applied Cognitive Psychology, 26*(5), 785–795. https://doi.org/10.1002/acp.2860

Harris, M. R., Fein, E. C., & Machin, M. A. (2022). A systematic review of multilevel influenced risk-taking in helicopter and small airplane normal operations. *Frontiers in Public Health, 10,* 823276. https://doi.org/10.3389/fpubh.2022.823276

Helmreich, R. L. (1984). Cockpit management attitudes. *Human Factors, 26*(5), 583–589. https://doi.org/10.1177/001872088402600510

Hunter, D. R. (2002). *Risk Perception and Risk Tolerance in Aircraft Pilots: (646002007-001)* [Data set]. American Psychological Association. https://doi.org/10.1037/e646002007-001

Hunter, D. R. (2005). Measurement of hazardous attitudes among pilots. *The International Journal of Aviation Psychology, 15*(1), 23–43. https://doi.org/10.1207/s15327108ijap1501_2

King, R. E., Retzlaff, P. D., Barto, E. B., Ree, M. J., & Teachout, M. S. (2012). *Pilot Personality and Training Outcomes* (p. 0025). (Technical Report No. AFRL-SA-WP-TR-2012-0013). Wright-Patterson AFB, OH: US School of Aerospace Medicine. www.dtic.mil/cgi-bin/GetTRDoc

Korte, R. F. (2007). A review of social identity theory with implications for training and development. *Journal of European Industrial Training, 31*(3), 166–180. https://doi.org/10.1108/03090590710739250

Li, T., & Zhao, Y. (2012). Help less or help more—perceived intergroup threat and out-group helping. *International Journal of Psychological Studies, 4*(10), 5539.

Maglio, M. A., Scott, C., Davis, A. L., Allen, J., & Taylor, J. A. (2016). Situational pressures that influence firefighters' decision making about personal protective equipment: A qualitative analysis. *American Journal of Health Behavior, 40*(5), 555–567. https://doi.org/10.5993/AJHB.40.5.2

Makarowski, R., Makarowski, P., Smolicz, T., & Plopa, M. (2016). Risk profiling of airline pilots: Experience, temperamental traits and aggression. *Journal of Air Transport Management, 57,* 298–305. https://doi.org/10.1016/j.jairtraman.2016.08.013

Michalski, D. J., & Bearman, C. (2014). Factors affecting the decision making of pilots who fly in Outback Australia. *Safety Science, 68,* 288–293. https://doi.org/10.1016/j.ssci.2014.03.005

Murray, S. R. (1999). FACE: Fear of loss of face and the five hazardous attitudes concept. *The International Journal of Aviation Psychology, 9*(4), 403–411. https://doi.org/10.1207/s15327108ijap0904_6

National Transportation Safety Board. (2020). *Safety Recommendation Report: Revise Processes to Implement Safety Enhancements for Alaska Aviation Operations.* www.ntsb.gov/investigations/AccidentReports/Reports/ASR2002.pdf

Neff, P. S. (2022). The five hazardous attitudes, a subset of complacency. *International Journal of Aviation, Aeronautics, and Aerospace, 9*(1). https://doi.org/10.15394/ijaaa.2022.1677

Noort, M. C., Reader, T. W., & Gillespie, A. (2021). Safety voice and safety listening during aviation accidents: Cockpit voice recordings reveal that speaking-up to power is not enough. *Safety Science, 139,* Article 105260. https://doi.org/10.1016/j.ssci.2021.105260

O'Bryan, K. M. (2011). Freud's ego in the cockpit. *Journal of Aviation/Aerospace Education & Research,* 21(1). http://commons.erau.edu/jaaer/vol21/iss1/5

Orasanu, J., Fischer, U., & Davison, J. (2002). Risk perception: A critical element of aviation safety. *IFAC Proceedings,* 35(1), 49–58. https://doi.org/10.3182/20020721-6-ES-1901.01639

Paletz, S. B. F., Bearman, C., Orasanu, J., & Holbrook, J. (2009). Socializing the human factors analysis and classification system: Incorporating social psychological phenomena into a human factors error classification system. *Human Factors,* 51(4), 435–445. https://doi.org/10.1177/0018720809343588

Pascual, A., Guéguen, N., Pujos, S., & Felonneau, M. L. (2013). Foot-in-the-door and problematic requests: A field experiment. *Social Influence,* 8(1), 46–53. https://doi.org/10.1080/15534510.2012.696038

Personality and Attitude Effects. (2021, April 27). Retrieved (05/07/2024). https://socialsci.libretexts.org/@go/page/15478

Ryckman, R. (2013). *Theories of Personality* (10th ed.). San Francisco: Cengage Advantage Books.

Schunk, D. H., & DiBenedetto, M. K. (2020). Motivation and social cognitive theory. *Contemporary Educational Psychology, 60,* Article 101832. https://doi.org/10.1016/j.cedpsych.2019.101832

Stajkovic, A., & Luthans, F. (2002). Social cognitive theory and self-efficacy: Implications for motivation theory and practice. In R. M. Steers, L. W. Porter, & G. A., Bigley (Eds.), *Motivation and Work Behavior* (7th ed., pp. 126–140). https://www.researchgate.net/publication/258995495_Social_cognitive_theory_and_self-efficacy_Implications_for_motivation_theory_and_practice

Stajkovic, A., & Sergent, K. (2019). *Social Cognitive Theory.* Obo in Management. https://doi.org/10.1093/obo/9780199846740-0169

Stewart, J. E., II. (2006). *Locus of Control, Attribution Theory, and the "Five Deadly Sins" of Aviation.* (Technical Report No. 1182). U.S. Army Research Institute for the Behavioral and Social Sciences. https://apps.dtic.mil/sti/citations/ADA452056

Sutin, A. R., Costa, P. T., Jr., Miech, R., & Eaton, W. W. (2009). Personality and career success: Concurrent and longitudinal relations. *European Journal of Personality, 23*(2), 71–84. https://doi.org/10.1002/per.704

Tajfel, H., & Turner, J. C. (1979). An integrative theory of intergroup conflict. In W. G. Austin, & S. Worchel (Eds.), *The Social Psychology of Intergroup Relations* (pp. 33–48). Monterey, CA: Brooks/Cole.

Velazquez, J. (2018). The presence of behavioral traps in U.S. airline accidents: A qualitative analysis. *Safety (Basel), 4*(1), 2. https://doi.org/10.3390/safety4010002

Wagstaff, P. (2016, February 16). *Ego. What Drives Your Decision-Making*? www.planeandpilotmag.com/article/ego/

Wiggins, M., Hunter, D., O'Hare, D., & Martinussen, M. (2012). Characteristics of pilots who report deliberate versus inadvertent visual flight into instrument meteorological conditions. *Safety Science—SAF SCI.* https://doi.org/10.1016/j.ssci.2011.10.010

Wood, R., & Bandura, A. (1989). Social cognitive theory of organizational management. *Academy of Management Review, 14*(3), 361–384. https://doi.org/10.5465/AMR.1989.4279067

CHAPTER 4

Aircraft Owners and Pilots Association. (2018, December 18). *Advocacy: Emergency Locator Transmitter.* Retrieved (5/8/2024). www.aopa.org/advocacy/aircraft/aircraft-operations/emergency-locator-transmitters

Alexander, D. J. (2016). The genesis of crew resource management: The NASA experience. In L. Gillman, S. Widder, M. D. Blaivas, & D. Karakitsos (Eds.), *Trauma Team Dynamics.* Cham: Springer. https://doi.org/10.1007/978-3-319-16586-8_1

Boeing. (2023, September). *Statistical Summary of Commercial Jet Airplane Accidents Worldwide Operations (19592022).* www.faa.gov/sites/faa.gov/files/2023-10/statsum_summary_2022.pdf

Farago, P., Shuffler, M. L., & Salas, E. (2019). Chapter 9: The design, delivery, and evaluation of crew resource management training. In B. G. Kanki, J. Anca, & T. R. Chidester (Eds.), *Crew Resource Management* (3rd ed., pp. 251–282). Academic Press. https://doi.org/10.1016/B978-0-12-812995-1.00009-9.

Federal Aviation Administration. (2021, November 17). *Safety Management Systems (SMS) Explained.* Retrieved (5/8/2024). www.faa.gov/about/initiatives/sms/explained/basis

Federal Aviation Administration. (2023). *Information to Operators.* www.faa.gov/other_visit/aviation_industry/airline_operators/airline_safety/info/all_infos/info23003

Friedman-Berg, F., Allendoerfer, K., & Pai, S. (2008). Nuisance alerts in operational ATC environments: Classification and frequencies. In *Proceedings of the Human Factors and Ergonomics Society Annual Meeting.* https://doi.org/10.1177/154193120805200123

Helmreich, R. L., & Foushee, H. C. (2019). Chapter 1: Why CRM? empirical and theoretical bases of human factors training. In B. G. Kanki, J. Anca, & T. R. Chidester (Eds.), *Crew Resource Management* (3rd ed., pp. 3–52). Academic Press. https://doi.org/10.1016/B978-0-12-812995-1.00001-4

International Air Transportation Association. (2018). *IATA Controlled Flight into Terrain Accident Analysis Report, 2008–2017 Data.* www.iata.org/contentassets/

International Civil Aviation Organization. (2018). *Safety Management Manual. Doc. 9859* (4th ed.). Montreal, Canada. https://skybrary.aero/sites/default/files/bookshelf/5863.pdf

Michalski, D. J., & Bearman, C. (2014). Factors affecting the decision making of pilots who fly in Outback Australia. *Safety Science, 68,* 288–293. https://doi.org/10.1016/j.ssci.2014.03.005

National Transportation Safety Board. (2017a). *Collision with Terrain, Promech Air, Inc. de Havilland DHC-3, N270PA, Ketchikan, AK.* June 25, 2015 (Aircraft Accident Report No. NTSB/AAR-17/02). www.ntsb.gov/investigations/AccidentReports/Reports/AAR1702.pdf

National Transportation Safety Board. (2017b). *Investigative Hearing Flight 3153 Anchorage, Alaska Aug 17, 2017 pt. 1.* [Video]. YouTube. https://www.youtube.com/watch?v=ZrUYwV4oECI&t=1s

National Transportation Safety Board. (2018a). *Collision with Terrain, Hageland Aviation Services, Inc. dba Ravn Connect Flight 3153, Cessna 208B, N208SD, Togiak, Alaska.* October 2, 2016 (Report No. NTSB/AAR-18/02).

National Transportation Safety Board. (2018b). *Collision with Terrain, Hageland Aviation Services, Inc. dba Ravn Connect Flight 3153, Cessna 208B, N208SD, Togiak, Alaska. Docket: Exhibit 14C—Human Performance Attachment 2—Interview Transcripts.* (Report No. NTSB/AAR-18-02).

National Transportation Safety Board. (2020a). *Improve the Safety of Part 135 Aircraft Flight Operations.* Retrieved (5/10/2024). www.ntsb.gov/Advocacy/mwl/Documents/2019-20/2019-20-MWL3-Part135.pdf

National Transportation Safety Board. (2020b). *Safety Recommendation Report: Revise Processes to Implement Safety Enhancements for Alaska Aviation Operations.* www.ntsb.gov/investigations/AccidentReports/Reports/ASR2002.pdf

National Transportation Safety Board. (2021). *Midair Collision over George Inlet, de Havilland DHC-2, N952DB, and de Havilland DHC-3, N959PA, Ketchikan, Alaska.* May 13, 2019 (Report No. NTSB/AAR-21/04). https://data.ntsb.gov/Docket?ProjectID=99423

Noort, M. C., Reader, T. W., & Gillespie, A. (2021). Safety voice and safety listening during aviation accidents: Cockpit voice recordings reveal that speaking-up to power is not enough. *Safety Science, 139,* Article 105260. https://doi.org/10.1016/j.ssci.2021.105260

Ruskin, K., Corvin, J., Stephen Rice, S., Richards, G., Winter, S. R., & Ruskin, A. C. (2021). Alarms, alerts, and warnings in air traffic control: An analysis of reports from the aviation safety reporting system. *Transportation Research Interdisciplinary Perspectives, 12,* 100502. https://doi.org/10.1016/j.trip.2021.100502

Sumwalt, R. (2019, October 10). *Roundtable Discussion Yields Key Insights, Critical Actions Needed for Improving Safety of Part 135 Flight Operations in Alaska.* NTSB Safety Compass. https://safetycompass.wordpress.com/2019/10/10/roundtable-discussion-yields-key-insights-critical-actions-needed-for-improving-safety-of-part-135-flight-operations-in-alaska/

Tomaszewska, J., Woch, M., Krzyszkowski, J., Zieja, M., & Grzesik, N. (2020). Analysis of the causes of aviation accidents during controlled flight into terrain. *Proceedings of the 29th European Safety and Reliability Conference (ESREL),* 1574–1578. https://doi.org/10.3850/978-981-14-8593-0_3947-cd

Transport Canada. (2019, July). *Evaluation of Safety Management Systems in Civil Aviation.* https://tc.canada.ca/sites/default/files/2021-02/evaluation-safety-management-systems-civil-aviation.pdf

Wiggins, M., Hunter, D., O'Hare, D., & Martinussen, M. (2012). Characteristics of pilots who report deliberate versus inadvertent visual flight into instrument meteorological conditions. *Safety Science—SAF SCI.* https://doi.org/10.1016/j.ssci.2011.10.010

CHAPTER 5

Atkins, D. (2021). *Understanding Decision-Making Styles of Part 135 Pilots in Remote Settings and Their Impact on Aviation Safety: A Qualitative Enquiry of Alaskan Operational Environments.* Theses and Dissertations, 4156. https://commons.und.edu/theses/4156

Saldaña, J., & Omasta, M. (2018). *Qualitative Research: Analyzing Life.* Thousand Oaks, California: SAGE.

CHAPTER 6

Carroll, M. (2023). Decision making in aviation. In J. R. Keebler, E. H. Lazzara, K. A. Wilson, & E. L. Blickensderfer (Eds.), *Human Factors in Aviation and Aerospace* (3rd ed., pp. 563–588). Academic Press. https://doi.org/10.1016/B978-0-12-420139-2.00016-2

Chira, T. I., & Adams, M. (2011). Behavioral bias within the decision-making process. *Journal of Business and Economics Research*, *6*(8). https://doi.org/10.19030/jber.v6i8.2456

Department of the Air Force. (2019). *Air Force Mentoring Program*. Department of the Air Force. https://afrotc.virginia.edu/sites/afrotc/files/2023-05/afh36-2643.pdf

Fitzgibbons, A., Schutte, P., & David, D. (2000). *Pilot Personality Profile using the NEO-PI-R*. https://ntrs.nasa.gov/citations/20000105204

Wickens, C. D., Helton, W. S., Hollands, J. G., & Banbury, S. (2021). *Engineering Psychology and Human Performance* (5th ed.). Routledge. https://doi.org/10.4324/9781003177616

CHAPTER 7

Bearman, C., Paletz, S. B. F., Orasanu, J., & Brooks, B. P. (2009). Organizational pressures and mitigating strategies in small commercial aviation: Findings from Alaska. *Aviation, Space, and Environmental Medicine, 80*(12), 1055–1058. http://dx.doi.org/10.3357/ASEM.2590.2009

Federal Aviation Administration. (2021). *FAA Alaska Aviation Safety Initiative FY21 Final Report*. www.faa.gov/sites/faa.gov/files/2021-10/FAA%20Alaska%20Safety%20Initiative%20Report%20-%20final.pdf

Federal Aviation Administration. (2024, January 17). *Part 139 Airport Certification*. www.faa.gov/airports/airport_safety/part139_cert

Michalski, D. J., & Bearman, C. (2014). Factors affecting the decision making of pilots who fly in Outback Australia. *Safety Science, 68,* 288–293. https://doi.org/10.1016/j.ssci.2014.03.005

Wood, R., & Bandura, A. (1989). Social cognitive theory of organizational management. *Academy of Management Review, 14*(3), 361–384. https://doi.org/10.5465/AMR.1989.4279067

Index

For Product Safety Concerns and Information please contact our EU
representative GPSR@taylorandfrancis.com
Taylor & Francis Verlag GmbH, Kaufingerstraße 24, 80331 München, Germany

www.ingramcontent.com/pod-product-compliance
Ingram Content Group UK Ltd.
Pitfield, Milton Keynes, MK11 3LW, UK
UKHW022158190926
475702UK00003B/83
9781032674742